Reasoning with the Infinite

Reasoning with the Infinite

From the Closed World to the Mathematical Universe

Michel Blay

Translated by M. B. DeBevoise

THE UNIVERSITY OF CHICAGO PRESS

CHICAGO AND LONDON

Michel Blay is director of research at the Centre National de la Recherche Scientifique, co-director of the Centre International de Synthese, and editor-in-chief of *Revue d'Histoire des Sciences*. Among his other books are *Les Figures de l'Arc-en-Ciel* and *La Naissance de la Mécanique Analytique*.

The University of Chicago Press, Chicago 60637
The University of Chicago Press, Ltd., London
© 1998 by The University of Chicago
All rights reserved. Published 1998
Printed in the United States of America
07 06 05 04 03 02 01 00 99 98   5 4 3 2 1

ISBN (cloth): 0-226-05834-4

Originally published as *Les raisons de l'infini: Du monde clos à l'univers mathématique*, © Éditions Gallimard, 1993.

Library of Congress Cataloging-in-Publication Data

Blay, Michel, 1948–
    [Raisons de l'infini. English]
    Reasoning with the infinite : from the closed world to the
mathematical universe / Michel Blay ; translated by M. B. DeBevoise.
        p.   cm.
    Includes bibliographical references and index.
    ISBN 0-226-05834-4 (alk. paper)
    1. Motion.   2. Infinite.   3. Mathematics—Philosophy.
    4. Mathematical physics.   I. Title.
    QC133.B53   1998
    530.15—dc21   97-49423
                                                    CIP

Anaximander [. . .] said that the Unlimited is the principle and the element of things that are, being besides the first to use the term ["]principle["]. He says that it is neither water, nor any other of what are said to be elements, but that it is some other unlimited nature from which are engendered the heavens and all the worlds that are found in them [. . .].

Simplicius, *In Aristotelis physicorum libros commentaria* (Dumont, Delattre, and Poirier, eds., *Les Présocratiques*, 26–27)

CONTENTS

The French title of Michel Blay's book, *Les raisons de l'infini,* bears a dual sense that cannot be carried over into English without taking certain liberties. The *raisons* of infinity with which it is concerned are not only the mathematical ratios that the new calculus of the late seventeenth century devised novel methods for handling with reference to physical motion but also the reasons that would justify such methods and give them a logically satisfactory foundation. Indeed it is the attempt, ultimately abandoned by the end of the following century, to make sense of the infant science of mathematical physics, to find significance in it comparable to that which permitted Galileo and Descartes to claim on behalf of the older physics fundamental insight into the nature of the world, that furnishes the author with his main theme. On this account the book might be called, somewhat misleadingly perhaps, *On the Meaning and Uses of the Infinite.*

In the interests of avoiding inadvertent anachronism, a few terms in the French edition have been rendered differently depending on the period to which they refer. Thus, for example, in connection with Huygens's work, *pesanteur* has been consistently translated as "weight" rather than "gravity" in order to preserve the Cartesian spirit of these texts (as against later Newtonian usage). On the other hand, I have for reasons of style tended somewhat unhistorically to treat "speed" and "velocity" as synonymous throughout the book, while on balance preferring the former for authors prior to 1700. Though, strictly speaking, the term "velocity" does not enter into mathematical physics until the nineteenth century, the concept it expresses was understood by Galileo, for instance, and to the extent that the motions considered in the seventeenth century were usually linear, circular, or elliptical, direction was implicit.

In the case of quoted passages for which no standard English translation exists, as occurs with authors such as Fontenelle, and not infrequently with Huygens and Leibniz as well, I have tried to preserve as far as possible the archaic air and cadence of these writers rather than to modernize their style, in order to enable the reader to have a feeling for how differently they conceived and formulated the problems that were at issue then than we do today. In this way the English-speaking student, who enjoys the same advantage with regard to Hobbes and Newton and Berkeley that his or her French counterpart enjoys reading native authors, or foreign authors of the period who wrote in French, may have a roughly similar sense of how Leibniz and Varignon and the Bernoullis sounded to their contemporaries. Achieving this effect means translating more, rather than less, literally, altering punctuation as little as possible, retaining ampersands and other distinctive elements of the original typography, and inserting bracketed interpolations where necessary for the sake of sense and flow. Where it is helpful to indicate modern formulations, this has been done in the notes.

Finally, a considerable amount of bibliographical information not found in the original edition of the book has been included in the English edition. I am grateful to Susan Abrams and her assistants at the University of Chicago Press, particularly Zarena Aslami, for their help in locating translated works; to Dr. Trevor C. Lipscombe for guidance on technical points; to Pamela J. Bruton, whose expert copyediting saved me from error more than once; and, not least, to the author for his patience in responding to queries and his painstaking review of the final draft. Whatever mistakes remain are my responsibility alone.

M. B. DeBevoise

Introduction

Galileo, in *Il Saggiatore* (1623), advanced a thesis that was to become famous: "Philosophy is written in this grand book—I mean the universe—which stands continually open to our gaze, but it cannot be understood unless one first learns to comprehend the language and interpret the characters in which it is written. It is written in the language of mathematics [. . .]."[1] The universe, Galileo claimed, is written in mathematical language. Which mathematical language, and which characters, did he have in mind?

The rest of the passage furnishes a clue: "It is written in the language of mathematics, and its characters are triangles, circles, and other geometrical figures, without which it is humanly impossible to understand a single word of it; without these, one is wandering about in a dark labyrinth."[2] The same ideas were taken up again some years later in a letter Galileo wrote to Fortunio Licetti dated January 1641: "But I truly believe the book of philosophy to be that which stands perpetually open before our eyes, though since it is written in characters different from those of our alphabet it cannot be read by everyone; and the characters of such a book are triangles, squares, circles, spheres, cones, pyramids, and other mathematical figures, most apt for such reading."[3] The book of the world seems to be written, then, essentially on the basis of geometrical figures, or "characters," such as the triangle, the circle, the sphere, and so on—these construed in terms of a Euclidean grammar.

Beyond the polemical aspect of these passages, the project Galileo proposed was extraordinary in its daring, for at the beginning of the seventeenth century, examples that might have been looked to for evidence of such an assertion, in its universal import, were few indeed.

Archimedes had already mobilized the resources of mathematics for the purpose of working out physical problems and, in two treatises, *On the Equilibrium of Planes* and *On Floating Bodies,* had put forth a set of propositions inspired by the Euclidean deductive model.

One encounters a very similar situation in another field of scientific knowledge: geometrical optics. Here the *Optics* of Euclid and Ptolemy made it possible to assemble a set of well-ordered propositions, to which Kepler's remarkable *Ad Vitellionem Paralipomena* (1604) was to find itself indebted on a good many points.

Finally, one may also recall the work of the Oxford School and of Nicolas Oresme in France, in the fourteenth century, on the kinematic implications of the doctrine of the "latitude of forms."

These various theoretical constructions, in which Euclidean geometry played a determining role, appear however only as particular applications of mathematics and not as results of a process arising from a truly universal design in which geometry manifests itself as the means permitting human thought to penetrate into the nature of things, or at least, to repeat Galileo's figure more precisely, as the means by which it becomes possible to understand the language the Creator used to express himself.

Galileo's project was, in its general form, Descartes's project as well—in the sense, to put the point briefly, that Euclidean geometry equally enabled the Cartesian program to grasp that "which constitutes the nature of bodies":

> 4. The nature of body consists not in weight, hardness, colour, or the like, but simply in extension.
>
> If we do this, we shall perceive that the nature of matter, or body considered in general, consists not in its being something which is hard or heavy or coloured, or which affects the senses in any way, but simply in its being something which is extended in length, breadth and depth. For as regards hardness, our sensation tells us no more than that the parts of a hard body resist the motion of our hands when we come into contact with them. If, whenever our hands moved in a given direction, all the bodies in that area were to move away at the same speed as our approaching hands, we should never have any sensation of hardness. And since it is quite unintelligible to suppose that, if bodies did move away in this fashion, they would thereby lose their bodily nature, it follows that this nature cannot consist in hardness. By the same reasoning it can be shown that weight, colour,

and all other such qualities that are perceived by the senses as being in corporeal matter, can be removed from it, while the matter itself remains intact; it thus follows that its nature does not depend on any of these qualities.[4]

Thus, and without wishing to multiply examples and citations, it is plain that for Galileo and Descartes alike Euclidean geometry was not only the model of the deductive organization of all well-founded knowledge but also, and above all, the means by which the nature of things could really be grasped and comprehended.

If this ambition remains ours still today, in the form of mathematical physics, it is however no longer the same in its essential aim. The construction of a mathematical science of nature is conceived no longer as an enterprise that allows the nature of things to be penetrated via geometry but rather as an undertaking that consists at best in constructing a coherent system of axioms, principles, and concepts from which certain conclusions or deductions may be compared with experience.

Geometrization, which is to say the project of Galileo and Descartes in its initial ontologico-geometrical purpose, has been abandoned, and its place taken by what is usually called mathematization. This term is to be understood chiefly as indicating an approach whose object is to reconstruct the phenomena of nature within the domain of mathematical intelligibility in such a way that these phenomena find themselves governed by quantitative laws which can be exploited for the purpose of predicting the course of nature by means of mathematical reason, thus ensuring the domination of nature by mathematical reason. But this is not all: to mathematize this or that natural phenomenon also means presenting in an ordered fashion the set of theorems, propositions, and results that one has managed to establish. Through such a deductive form of organization, each proposition having been obtained on the basis of the preceding ones, clarification and methodical investigation of the fundamental properties of various phenomena become possible by deploying all the resources of the mathematical knowledge of a given period to this end.

From this perspective the first thoroughly finished example of the deductive ideal of the process of mathematization and, correspondingly, of the constitution of a mathematical physics, was Joseph-Louis Lagrange's *Mécanique analytique* (1788).

For the first time, the whole of mechanics (or, more precisely, "the theory of this Science, and the art of resolving the problems that relate to it") was in place, with the help of the algorithms of the differential calculus and in particular the application of equations involving partial derivatives, now reduced, as Lagrange wrote, "to general formulas whose simple development gives all the necessary formulas for the solution of each problem [. . .]. The methods that I lay out require neither constructions, nor geometrical or mechanical arguments, but only algebraic operations, subject to a regular and uniform procedure. Those who love Analysis will take pleasure in seeing mechanics become a new branch of it, and will be grateful to me for thus having extended its domain."[5]

The development of physics thus proceeded by means of a process of mathematization intimately associating mathematics and physics in such a way that the discourse of physics became inseparable from its mathematical form, so that there was now a cosubstantiality, as it were, between physics and mathematics, or, as Gaston Bachelard put the point, "[. . .] the hypotheses of physics are formulated mathematically. Scientific hypotheses are henceforth inseparable from their mathematical form: they are truly mathematical thoughts."[6]

Thus the mathematization of natural phenomena, while retaining the initial insistence on the requirements of deductive organization inherited from the Euclidean geometric model, put aside any meaningful ontological aims it might have had.

How and why was the initial project abandoned, at least in its full range and ambition? More precisely, why was geometrization transformed into what is commonly called mathematization?

In order to answer these questions, we have to reexamine the first stages of the process of geometrization. We have already seen, in both the passage quoted from Galileo's *Saggiatore* and in the second part of Descartes's *Principia*, that geometry manifested itself as the means by which human thought can enter into the nature of things or, at least, as the means by which it becomes possible to understand the language in which the Creator deigned to express himself. Now, it may very roughly be said that at this time the process of geometrization appeared unable to be carried out within the sphere of Euclidean geometry and Archimedeo-Euclidean approaches alone. In fact, in the case of the geometrization of motion—the veritable touchstone for geometrization in the seventeenth century—questions immediately arose of extreme difficulty, questions over which the shadow of Zeno of Elea had hovered for more than two millennia: In what do the beginning and end of motion consist? How are they to be accounted for geometrically? How is the continuity of motion to be grasped? Is motion actually continuous, or is it rather a combination of motion and rest? How is the sum of all the velocities (contained in motion) to be understood, and so on?

All these questions had one thing in common: answering them required dealing with the infinite, whether in the form of infinite series or infinite sums or else in the form of infinite division—problems that had come down through the centuries and to which the seventeenth century was at last to provide answers. What was at stake during this period was the pursuit of the very process of geometrization itself, and so the construction of a science of the nature of things as well.

In the first instance, then, to deal with the infinite—to construct a concept of infinity in mathematical terms—was a matter both (though it may have meant transgressing the usual and traditional laws of logic) of inquiring into indivisibles and the composition of the continuum and of attempting to get beyond Zeno's inevitable paradoxes (the "Achilles," the "Arrow," and the "Stadium"). Because the problem needing to be solved was a mathematical one, it was not enough, according to Edme Mariotte in his *Essai de logique* (1678), to be satisfied with experimental evidence, which settles nothing:

> Sophisms that rely on the infinite division of space proceed also from what cannot be understood[, that is,] the infinite, which has no determined meaning, on account of which one can scarcely find the solution to these Sophisms; but it suffices to give a contrary proof, easy to understand, as[, for example], if one wishes to prove that a man who runs twice as fast as another would never be able to catch up with him if the latter had a head start of a league; because during the time it takes the faster [runner] to cover this league, the other would have gone half a league further; and that when the faster one had made up this half-league, the other would have gone a quarter-league further, and so on to infinity; it must be replied that if the faster one does a league in an hour, and the other one a half-league in the same time, the first will have done three leagues in three hours, and the

other a league and a half, and that consequently the faster one will have overtaken him by half a league [after three hours]; this latter argument is clear, and the other is obscure.

And if to prove that there is no motion in nature, one says, "That which moves [either] moves in the place where it is, or in that where it no longer is, neither of which is possible; therefore no movement takes place": it must be replied that it is difficult or impossible to understand in detail how a body passes from one place to another, because the spaces [between them] are divisible to infinity, and one cannot understand infinity; but that one thing is very clear, and which no argument can destroy, [namely,] that bodies do change position.[7]

Beyond these already considerable mathematical difficulties, however, what the project of geometrization implied in its most profound sense was the introduction of the infinite into the world, and the affirmation of its presence in it, for it was only by virtue of this that the essential purpose of geometrization could be fully realized. Such a perspective allowed geometrization to regard the theses of Giordano Bruno as an inevitable consequence, particularly in the form given them in De l'infinito, universo e mondi (1584), and no longer as a matter of peremptory assertion: "There are no ends, boundaries, limits, or walls which can defraud or deprive us of the infinite multitude of things. Therefore the earth and the ocean thereof are fecund; therefore the sun's blaze is everlasting [. . .]. For from infinity is born an ever fresh abundance of matter."[8]

How could one conceive of a real infinite, present in the world, when it was exactly the conception of the infinite that was supposed to be reserved to the Creator of the world—when speaking the name of the infinite was reserved to God alone?[9]

The point of the enterprise that was implicitly to lead to the abandonment of the initial project of geometrization lies in the reply that unavoidably had to be given to this question. The position of Blaise Pascal (1623–1662) is particularly illuminating in this connection. Though he asserted that double infinity is encountered in everything, he simultaneously emphasized that this double infinity cannot be conceived by the human mind. Accordingly, human understanding cannot in fact wholly penetrate into the nature of things. The book of the universe is illegible.

Thus, in his short work De l'esprit géométrique (1657/58), Pascal writes:

[. . .] it [geometry] therefore supposes that we know what is understood by such words as *motion, number,* and *space;* and without spending time in useless definition, it discerns their nature and discovers their marvelous properties. These three things, which, according to the words *Deus fecit omnia in pondere, in numero, et mensura,* include the whole universe, and have a reciprocal and necessary connection. For we cannot imagine motion without something which moves; and since this thing is one, this unity is the source of all numbers. Finally, since motion cannot exist without space, we see these three things included in the first. Even time is likewise included, for motion and time are relative to each other inasmuch as speed and slowness, which are differences of motion, have a necessary relation to time. Thus there are properties common to all things, and the knowledge of them opens the mind to the greatest wonders of nature. The principal one includes the two infinities which are to be found in all things, infinite largeness and infinite smallness. For no matter how fast a motion may be, it is always possible to conceive one that is faster, and then to accelerate this one even more, and so on infinitely without ever attaining any motion so rapid that it cannot become more so. On the other hand, no matter how slow a motion may be, it can still be retarded, and this retarded motion can again be retarded, and so on infinitely without ever attaining such a degree of slowness that we cannot decrease the speed by an infinite number of gradations without reaching the point of rest. Likewise, no matter how large a number may be, we can always conceive a greater number, and then another to surpass this one, and so on infinitely without ever reaching one which cannot be increased. And conversely, no matter how small a number is, as for example, a hundredth or a ten-thousandth, we can still conceive of a smaller one, and so on infinitely, without arriving at zero or nothing. No matter how large a space is, we can imagine a larger one, and still a larger one than this, and so on infinitely, without ever arriving at one which could no longer be increased. And conversely, no matter how small a space may be, we can still think of a smaller one, and so on infinitely, without ever reaching one which is indivisible because it no longer has any extent.

The same applies to time. We can always conceive of a greater duration of time without final limit, and of a lesser duration without arriving at one moment and at an absolute cessation of duration. In short, this is the same as saying that no matter what motion, number, space, or time there may be, there is always one which is greater or less, so that they are all in progression between nothingness and infinity, always infinitely distant from these extremes.[10]

This passage, in asserting that "these three things [motion, number, and space] [. . .] include the whole universe" and are such that "the two infinities [. . .] are to be found in all things," therefore assigns the double infinity a place in the diversity of the universe, and so projects it into nature.

The same idea is forcefully repeated in fragment 199 of the *Pensées:* "When we know better, we understand that, since nature has engraved her own image, and that of her author on all things, they almost all share her double infinity."[11]

Contrary to what Fontenelle was later to affirm by his conception of "geometric infinity," the presence in the world of a double infinity did not imply for Pascal that such an infinity could be conceived. In this sense, Pascal did not go beyond the traditional argument that distinguished actual from potential infinity, or, following Cartesian terminology, between the infinite and the indefinite. The sole veritable infinity was that of God. Thus, writes Pascal in *De l'esprit géométrique,* "That is the admirable relation which nature has established between these things and the two marvelous infinities which she has offered to men, not to be conceived but to be admired."[12]

Thus too, to the same effect, he writes in fragment 199 of the *Pensées:*

> We naturally believe we are more capable of reaching the centre of things than of embracing their circumference, and the visible extent of the world is greater than we. But since we in our turn are greater than small things, we think we are more capable of mastering them, and yet it takes no less capacity to reach nothingness than the whole. In either case it takes an infinite capacity, and it seems to me that anyone who had understood the ultimate principles of things might also succeed in knowing infinity. One depends on the other, and one leads to the other. These extremes touch and join by going in opposite directions, and they meet in God and God alone.
>
> Let us then realize our limitations. We are something and we are not everything. Such being as we have conceals from us the knowledge of first principles, which arise from nothingness, and the smallness of our being hides infinity from our sight.[13]

Or again, as Pascal goes on to add later (in fragment 418 of the *Pensées*), more directly in connection with the number series, "We know that the infinite exists without knowing its nature, just as we know that it is untrue that numbers are finite. Thus it is true that there is an infinite number, but we do not know what it is."[14]

The world is infinite: it is everywhere shot through by the infinite; but the infinite is not of our world, in the sense that we can neither grasp it nor conceive it, only contemplate it. The construction of a mathematical concept of infinity that might also be a word belonging to the language of nature, fully comprehensible by human thought, lies therefore beyond our reach. Or, as Galileo had already expressed the point in the *Discorsi*: "[. . .] let us remember that we are among infinities and indivisibles, the former incomprehensible to our understanding by reason of their largeness, and the latter by their smallness. Yet we see that human reason does not want to abstain from giddying itself about them."[15] In the same spirit, as Descartes stressed in the *Principia*:

26. We should never enter into arguments about the infinite. Things in which we observe no limits—such as the extension of the world, the division of the parts of matter, the number of the stars, and so on—should instead be regarded as indefinite.

Thus we will never be involved in tiresome arguments about the infinite. For since we are finite, it would be absurd for us to determine anything concerning the infinite; for this would be to attempt to limit it and grasp it. So we shall not bother to reply to those who ask if half an infinite line would itself be infinite, or whether an infinite number is odd or even, and so on. It seems that nobody has any business to think about such matters unless he regards his own mind as infinite. For our part, in the case of anything in which, from some point of view, we are unable to discover a limit, we shall avoid asserting that it is infinite, and instead regard it as indefinite. There is, for example, no imaginable extension which is so great that we cannot understand the possibility of an even greater one; and so we shall describe the size of possible things as indefinite. Again, however many parts a body is divided into, each of the parts can still be understood to be divisible and so we shall hold that quantity is indefinitely divisible. Or again, no matter how great we imagine the number of stars to be, we still think that God could have created even more; and so we suppose the number of stars to be indefinite. And the same will apply in other cases.

27. The difference between the indefinite and the infinite.

Our reason for using the term "indefinite" rather than "infinite" in these cases is, in the first place, so as to reserve the term "infinite" for God alone. For in the case of God alone, not only do we fail to recognize any limits in any respect, but our understanding positively tells us that there are none. Secondly, in the case of other things, our understanding does not in the same way positively tell us that they lack limits in some respect; we merely acknowledge in a negative way that any limits which they may have cannot be discovered by us.[16]

The finitude of human thought, confronted with the infinitude of the Creator, prevented the process of geometrization from ever being able to be wholly realized. In so doing, it caused the project of geometrization to lose its meaning; it became impossible to read, and at the same time understand, infinity in nature—and therefore to fully fathom the nature of things. The fabric of nature had thus been torn, exposing a gap, as it were, between infinity, on the one hand—the discourse of the divine, its principles and foundations—and, on the other, human knowledge, which lacked any real depth, being subject to the reign of individual points of view and of ephemeral symbols.

Faced with this new order of meaning, the discourse of geometrization had no choice but to renounce all claims to foundational ontological purpose. And so Euclidean geometry, no longer being essential to the project, was abandoned—often reluctantly since, of course, abandoning it meant giving up any hope of attaining true knowledge. In place of geometrization was substituted mathematization. What was left, then, was only a well-constructed discourse which, in no longer speaking of the reality of things, in becoming detached from them, could freely employ the procedures of infinitesimal geometry and of the differential and integral calculus: because these procedures were now no more than methods, techniques, mere auxiliaries of calculation and investigation, the direct echo of which one could no longer claim to find in reality.[17]

The fact that only one other approach could be envisioned supplies the context in which the work of Fontenelle, in his *Elements de la géométrie de l'infini* (1727), is to be understood. The "elements" that Fontenelle claimed to construct were like those of Euclid, but they were elements that took into account the new calculus of infinity as part of an attempt at knowledge that carefully distinguished geometrical infinity from metaphysical infinity. If the world of Galileo and Descartes was written in Euclidean geometry, Fontenelle's world was to be written in the geometry of infinity:

"There is nothing which has been examined enough, and perhaps nothing will be examined enough. Everything in nature is infinite." "The infinite is everywhere in some manner or other, every finite thing resolves itself into [the] infinite."[18]

Despite his considerable efforts at rationalizing the infinite, as it were—in order to give meaning to the process of mathematization and to the construction of a mathematical physics through the attempt to anchor them in true knowledge—Fontenelle nonetheless failed. This failure, which to a large degree put the future of the new science at risk, operated on two levels: on the one hand, it affected the mathematics on which the new science was based, since the theoretical difficulties concerning the foundations of the calculus of the infinite (or, more precisely, after Leibniz, of the differential and integral calculus) were insurmountable at the beginning of the eighteenth century; and, on the other hand, it affected the larger interpretation of Fontenelle's work, which was not properly appreciated by his contemporaries, who concerned themselves only with the purely mathematical problems it raised.

The result was that geometrization once and for all gave way to mathematization. The success of the differential and integral calculus rapidly caused the old philosophical and theological issues to be forgotten, replacing them with wonder at the fecundity of the new technique, which supplied the spellbound analyst with so many novel rules and formulas and the like.

Thus Lagrange's work, which enjoyed an impressive success at the horizon of knowledge in the latter part of the eighteenth century, was to be accompanied by a positivist neglect of the question of what meaning could be assigned to mathematization and to what uses mathematical physics might be put.

The present essay therefore has the object of explaining, on the basis of the actual process of geometrization, the difficulties that arose from trying to take the infinite into account, as well as the techniques developed to resolve or to avoid them, insofar as the neglect of the question of meaning—which was associated with the hope for true knowledge—grew out of these very difficulties. Fontenelle's attempt to deal with them provides a posteriori evidence that this was the case.

The analysis given here is therefore situated between two particularly significant and limiting historical moments. The one, centered on Christiaan Huygens's work on the fall of heavy bodies, mainly occurred—so far as the substance of it was concerned—prior to the confrontation with infinity. Huygens, like Archimedes, though he took advantage of a particular method, nonetheless always chose to avoid speaking openly of the concept of infinity: an illusory prudence that sooner or later would have to be overcome. The other limiting moment, the subject of the epilogue, is centered on the various works of Bernard Le Bovier de Fontenelle—to the extent that these works, as has often been suggested, represent in spite of their failure a genuine attempt to take into account the notion of infinity as it was then being applied in mathematical physics, and to link it to the larger question of the meaning of this application and of the quest for true knowledge. Fontenelle's attempt stands finally as a symptom of a new style of thought that had not yet been fully worked out and that, ultimately, forces us to reconsider our own ontological assumptions.

ONE

Infinity Eliminated; or, Huygens's Theory of the Motion of Heavy Bodies

1. Establishing the General Fact of Gravity

If there was one book that marked the beginning of a new era in the science of motion in the seventeenth century, it was surely the one that Galileo (1564–1642) published in Italian—rather than Latin—at Leyden in 1638 under the title *Discorsi e dimostrazione matematiche intorno a due nuove scienze.*[1]

Galileo gave these *Discorsi* the form of a dialogue among three speakers: Salviati, friend and spokesman for Galileo; Sagredo, the honest man for whom demonstration and experiment take precedence over book learning; and Simplicio, the representative of official science and defender of orthodox scholasticism. The discussions among these three figures take place over the course of four days.

The first of the new sciences mentioned in the book's title—the study of the resistance of physical bodies—perfectly justified the claim of novelty made on its behalf. In fact, on the first day, Galileo lays out an approach to the problem of the cohesion of solids that leads him to introduce the hypothesis, among others, of a multitude of small intercalary voids that, in working against separation among the different parts of a body, tend to favor contact between them. He then, on the second day, organizes the problem of the resistance to rupture in the case of beams, prisms, and the like with reference to a lever, using the notion of momentum: "In such speculations I take as a known principle one which is demonstrated in mechanics about the properties of the rod which we call the lever: that in using a lever, the force is to the resistance in the inverse ratio of the distances from the fulcrum to the force and to the resistance."[2]

Galileo's treatment of these questions, to which he devoted himself as a result of conversations with technical experts in the shipyards of Venice, went substantially beyond the mathematical and technical knowledge of the time, particularly with regard to phenomena of elastic deformation. His work was to be considerably extended at the turn of the eighteenth century with the introduction of the methods of the differential calculus and the contributions of scientists such as Jacques Bernoulli (1654–1705), Pierre Varignon (1654–1722), and Antoine Parent (1666–1716). Their investigations were not to be fully integrated with each other, however, until the second half of the eighteenth century, with the development of a group of theoretical concepts that corresponds in its essential features to what is known today as the mechanics of continuous media, or continuum mechanics.[3]

It was therefore rather the second of the new sciences that was to rapidly ensure the success of Galileo's work. This was concerned with the study of both uniform and uniformly (or naturally) accelerated rectilinear motion, as well as with the composition of these two motions in trajectories described by projectiles.

For Galileo, in the *Discorsi*, it was a question not of tackling uniformly accelerated motion as a simple mathematical exercise but of resolving problems related to this kind of motion while simultaneously affirming that all bodies in natural free fall are impelled by such a motion. This assertion was far from self-evident. In fact, the diversity of forms exhibited by natural free motion made it difficult to imagine that a science of such motion might be possible at all. How could a single law of motion regulate both the fall of a fluttering feather and that of a swiftly dropping lead ball?

Galileo's essential contribution to the theory of the motion of heavy bodies resides in his discovery of just such a law. His approach, memorably analyzed by Maurice Clavelin in his *Philosophie naturelle de Galilée*,[4] consisted first of all, on the first day, in revealing by means of a passage to the limit—or rather, though recourse to experiment remained fundamental, in constructing— the general fact of the fall of heavy bodies, according to which bodies fall in a frictionless environment with the same speed:

> This seen, I say, I came to the opinion that if one were to remove entirely the resistance of the medium, all materials would descend with equal speed.[5]

And a bit further on:

> I repeat that my intention is to explain that the cause of diverse
> speeds in moveables of different heaviness is not that different
> heaviness at all, but depends on external events, particularly on
> the resistance of the medium, in such a way that by taking that
> away, all moveables would move at the same degrees of speed.[6]

It was on this basis, the general fact of gravity, that the development of a science of the motion of heavy bodies became possible.
As important as it was, the result concerning natural free motion
hardly provided all the elements necessary for arriving at a complete understanding of such motion. In particular, this result said
nothing about how an increase in speed comes about. As a proportion of space or of time, for example, does it increase uniformly
or not?

Galileo replies to these questions in the opening lines of the third
day of the *Discorsi,* in the section dealing with uniformly accelerated motion:

> On Naturally Accelerated Motion
> Those things that happen which relate to equable motion have
> been considered in the preceding book; next, accelerated motion
> is to be treated of.
>
> And first, it is appropriate to seek out and clarify the definition
> that best agrees with that [accelerated motion] which nature em-
> ploys. Not that there is anything wrong with inventing at plea-
> sure some kind of motion and theorizing about its consequent
> properties, in the way that some men have derived spiral and con-
> choidal lines from certain motions, though nature makes no use
> of these [paths]; and by pretending these, men have laudably
> demonstrated their essentials from assumptions [*ex suppositi-
> one*]. But since nature does employ a certain kind of acceleration
> for descending heavy things, we decided to look into their prop-
> erties so that we might be sure that the definition of accelerated
> motion which we are about to adduce agrees with the essence
> of naturally accelerated motion. And at length, after continual
> agitation of mind, we are confident that this has been found,
> chiefly for the very powerful reason that the essentials succes-
> sively demonstrated by us correspond to, and are seen to be in
> agreement with, that which physical experiments [*naturalia ex-
> perimenta*] show forth to the senses. Further, it as though we
> have been led by the hand to the investigation of naturally accel-
> erated motion by consideration of the custom and procedure of
> nature herself in all her other works, in the performance of which

she habitually employs the first, simplest, and easiest means. And indeed, no one of judgment believes that swimming or flying can be accomplished in a simpler or easier way than that which fish and birds employ by natural instinct.

Thus when I consider that a stone, falling from rest at some height, successively acquires new increments of speed, why should I not believe that those additions are made by the simplest and most evident rule? For if we look into this attentively, we can discover no simpler addition and increase than that which is added on always the same way. We easily understand that the closest affinity holds between time and motion, and thus equable and uniform motion is defined through uniformities of times and spaces; and indeed, we call movement equable when in equal times equal spaces are traversed. And by this same equality of parts of time, we can perceive the increase of swiftness to be made simply, conceiving mentally that this motion is uniformly and continually accelerated in the same way whenever, in any equal times, equal additions of swiftness are added on.

Thus, taking any equal particles of time whatever, from the first instant in which the moveable departs from rest and descent is begun, the degree of swiftness acquired in the first and second little parts of time [together] is double the degree that the moveable acquired in the first little part [of time]; and the degree that it gets in three little parts of time is triple; and in four, quadruple that same degree [acquired] in the first particle of time. So, for clearer understanding, if the moveable were to continue its motion at the degree of momentum [*momentum velocitatis*] of speed acquired in the first little part of time, and were to extend its motion successively and equably with that degree, this movement would be twice as slow as [that] at the degree of speed obtained in two little parts of time. And thus it is seen that we shall not depart from the correct rule if we assume that the intensification of speed [*intensionem velocitatis*] is made according to the extension of time [*fieri juxta temporis extensionem*]; from which the definition of the motion of which we are going to treat may be put thus:

[Definition]

I say that that motion is equably or uniformly accelerated which, abandoning rest, adds on to itself equal momenta of swiftness in equal times.[7]

This "admirable" and "extraordinarily complex" passage, as Clavelin has characterized it,[8] perfectly expresses the culmination of Galileo's attempt to establish the general fact of gravity in its full extent. The Galilean construction, once having deliberately made appeal to the principle of simplicity ("nature herself in all her other works [. . .] habitually employs the first, simplest, and easiest means"), therefore posited two things:

• on the one hand, that the independent variable (to use the modern term) to which the increase of speed in this motion is to be related is time and not space, as formerly had been supposed;[9] and,

• on the other hand, that naturally accelerated motion is also uniformly accelerated, which is to say that the increase in speed occurs as a simple proportion of the selected independent variable, time.

The third day of the *Discorsi* offers, then, a deductively formulated theory of naturally accelerated motion. Its best known theorems are the following:

Proposition 1. Theorem 1

The time in which a certain space is traversed by a moveable in uniformly accelerated movement from rest is equal to the time in which the same space would be traversed by the same moveable carried in uniform motion whose degree of speed is one-half the maximum and final degree of speed of the previous, uniformly accelerated, motion.[10]

It was this theorem of the "mean degree" that allowed a correspondence to be established between uniform motion and uniformly accelerated motion.[11] It also made it possible to establish the proportionality between space traversed and the square of the time, which is the object of the second theorem:

Proposition 2. Theorem 2

If a moveable descends from rest in uniformly accelerated motion, the spaces run through in any times whatever are to each other as the duplicate ratio of their times; that is, as are the squares of those times.[12]

Here we limit ourselves to stating these theorems without attempting to analyze their demonstrations, for we shall take them up again in a wider perspective in chapter 3.[13]

Up until this point, on the third day of the *Discorsi*, only natural free motion (that is, vertical motion) has been considered. In the third theorem, Galileo contemplates the case of a moving body placed on an inclined plane:[14]

Proposition 3. Theorem 3

If the same moveable is carried from rest on an inclined plane, and also along a vertical of the same height, the times of the movements will be to one another as the lengths of the plane and the vertical.[15]

He arrives at this result by applying the first theorem in conjunction with the principle formulated a few pages earlier, according to which "the degrees of speed that a single moveable acquires on differently inclined planes are equal, provided that the heights of those planes are equal."[16]

This third theorem made it possible, in fact, to compare the motions produced following oblique and vertical trajectories. As a result, it constituted a very important extension of the theory of the motion of heavy bodies and prepared the way for the study of the curvilinear trajectories described by such bodies, to the extent that it was justified to treat a curved line as a succession of small rectilinear segments.

2. Mathematical Speculations about Curvilinear Falls

Galileo's results were very quickly assimilated by Christiaan Huygens (1629–1695), as various manuscripts—mainly from the period 1657–1659—testify.[17] In particular, he reworked the Galilean principle of the equality of final speeds by relating it to what is often called Torricelli's principle.[18] From Galileo's principle, taken together with the last formula in Torricelli's *De motu gravium naturaliter descendentium et projectorum* (1644), it follows that the center of gravity of a system cannot be raised independently of some external force:

> We will posit in principle that two heavy bodies, bound together, cannot be moved of themselves, unless their common center of gravity falls.
>
> In fact, when two heavy bodies are bound together in such a way that the motion of the one entrains that of the other, whether that link be produced by the intermediary of a scale or of a pulley or of any other mechanism, these two bodies will behave as a single body made up of two parts; but such a heavy body will never put itself in motion, unless its center of gravity falls. Now therefore, if it is constituted in such a way that its center of gravity can in no way fall, the body will assuredly remain at rest in the position that it occupies; moreover, in fact, it would move

itself in vain, for it would follow a horizontal motion that tends not in the least downward.[19]

Torricelli, anticipating Huygens's work of some years later, applied this principle to the case of the motion of heavy bodies along inclined planes and demonstrated in a number of propositions the classical results concerning this question.[20]

In the manuscript dating from 1657 reproduced here, Huygens considered the horizontal line FD, the inclined plane BC, and the vertical line AB. If a moving body falls along the inclined plane BC, acquiring at B a degree of speed *(gradus velocitatis)* less than that which would have been acquired in falling from A to B, this body would however have acquired a degree of speed sufficient to permit it to climb back up as far as G such that, conversely, if a body falls from G to B it would thus climb back up as far as C. Thus a falling body could climb back up to a height higher than that from which it initially fell, which was contrary to the mechanical principle *("quod est contra principia Mechanices")* of Torricelli. As a result, a body falling from C must arrive at B with the same speed as that which it acquired in descending, in free fall, from A to B. This result can obviously be generalized to all inclined planes.[21]

Huygens's success in explaining motion along inclined planes immediately led him, as the same page of the manuscript also shows, to take an interest in motion along curvilinear trajectories. In fact, a curved line can be treated as a succession of small rectilinear segments, as so many small inclined planes, each one possessing, of course, a different degree of inclination with respect to a horizontal or vertical line.

During December 1659, in a series of passages that were likewise to remain in manuscript form, Huygens arrived at his celebrated results on the isochronism of cycloidal fall:[22] the time of fall from a physical point along any arc of the cycloid is constant when the point departs from rest, and all the arcs, no matter what their length, terminate at the vertex of the cycloid.[23] Huygens attached the greatest importance to this result and to its technical implications. Evidence of this may be found, for example, in the passage of the letter dated 22 January 1666 to Ismaël Boulliau (1605–1694) in which he asserts that this "fine invention [. . .] is the principal fruit that one could have hoped for from the science of accelerated motion, which Galileo had the honor of being the first to treat."[24]

Christiaan Huygens, manuscript of 1657 on motion along inclined planes. University of Leyden Library (*Hug. 10, fol. 84v*).

Huygens's results, beyond their theoretical importance, in fact rapidly found technical applications of the greatest interest for the construction of clocks: the curved blades between which a pendulum must oscillate needed to be given a cycloidal form to render the period of the oscillations theoretically independent of their amplitude.

It is therefore not on the evident importance of Huygens's result that we wish now to linger but rather on the way in which this result had been obtained in 1659 as an extension of Galileo's work, which relied on mathematical risk taking and innovation.

Huygens broached the quite tricky question of motions following curvilinear trajectories in a manuscript of December 1659—or, more exactly, in a series of manuscripts, the first of which bears the date 1 December 1659.[25]

The point of departure for Huygens's work grew out of his thinking and experiments on pendular motions, that is to say, as the opening lines of the manuscript put it, on the comparison between the time taken by a physical point to describe a very small pendular oscillation and that taken by the same physical point to fall vertically from an equivalent height.[26]

It was therefore research on motion following a circular arc as well as on small pendular oscillations that seems to have led Huygens to his results on the isochronism of falls in the cycloid.

Huygens considered therefore a pendulum whose extremity is treated as a physical point subject to the action of gravity. This extremity, departing from an arbitrarily chosen point K, describes a circular arc KEZ. Huygens then proposed comparing the motion along KEZ with the motion of a free-falling body, descending from the same height, along the line AZ. In fact, by using the first theorem (theorem of the mean degree) of the third day of the *Discorsi,* Huygens was to compare the motion along the circular arc KEZ, not with a uniformly accelerated motion, but with a uniform motion whose constant speed is equal to that reached by the body falling freely from A to Z. Let v_f be this final speed. It is important to recall here that in such a uniform motion the distance traversed, in the same time, is double that traversed through a motion in free fall, whose final speed is equal to the constant speed of uniform motion. As a result, a factor ½ was to appear in the last steps of Huygens's demonstration.[27]

Christiaan Huygens, manuscript of 1 December 1659 on motions following curvilinear trajectories. University of Leyden Library (*Hug. 26, fol. 72r*).

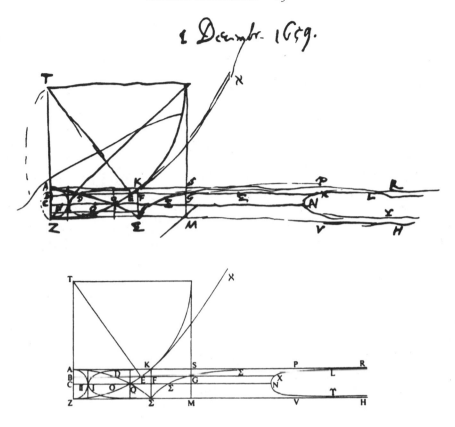

Christiaan Huygens, enlargement of head of 1 December 1659 manuscript with mechanical rendering.

Accordingly, Huygens compared the times of fall through the infinitely small parts of the arc KEZ and along the vertical AZ. Let (E) and (B) be these infinitely small parts *("particulam E," "particulam B")*, (B) being defined as the projection of (E) on the vertical AZ. Let us call $t_{(E)}$ the time of fall along (E) of a physical point departing from K without initial speed and $t_{(B)}$ the time it takes to traverse (B) with a uniform motion whose constant speed is equal to the final speed v_f of the motion in free fall from A to Z.

The infinitely small part (E) of the circular arc can be treated as an infinitely small rectilinear segment, with the result that the right triangles having sides (E) and (B) and vertices T and E are similar. From this it immediately follows that

$$\frac{(E)}{(B)} = \frac{TE}{BE} \tag{1}$$

Then, by construction, TE = GB, whence[28]

$$\frac{(E)}{(B)} = \frac{GB}{BE}$$

Moreover, the Galilean law of falling bodies stipulates that the spaces traversed are as the squares of the times, as a result of which the speeds of a physical point falling along AZ can be represented by the parabola ADΣ whose apex is in A. Thus the speed at B, for example, is expressed (or rather represented) by BD, and that at Z is expressed (or represented) by ZΣ. The spaces are represented by AZ. It follows that the speed at each point E on the circular arc KEZ can be expressed or represented by an ordinate BD of the parabola ADΣ. In fact, by the law of falling bodies (v_E being the speed of a physical point in E describing KEZ and v_B being the speed in B of a physical point in free fall from A to Z),

$$\frac{AB}{AZ} = \frac{v_B{}^2}{v_Z{}^2}$$

Now, for an equal height of fall, the speeds attained are equal: $v_E = v_B$; in consequence,

$$\frac{AB}{AZ} = \frac{v_E{}^2}{v_Z{}^2}$$

Then, by the definition of the parabola,

$$\frac{AB}{AZ} = \frac{BD^2}{Z\Sigma^2}$$

whence

$$\frac{v_E}{v_Z} = \frac{BD}{Z\Sigma} \quad \text{and} \quad v_E = v_Z \cdot \frac{BD}{Z\Sigma}$$

Now, since v_Z and ZΣ are, by hypothesis, given, the speed v_E is proportional to BD. It is therefore possible to express (or represent) the speed at each point E of KEZ by an ordinate BD of the parabola ADΣ.

Supposing now that the motions along the infinitely small parts (E) and (B) occur with constant speeds,[29] it follows that

$$\frac{(E)}{(B)} = \frac{v_{(E)}}{v_{(B)}} \cdot \frac{t_{(E)}}{t_{(B)}} \qquad \text{or} \qquad \frac{t_{(E)}}{t_{(B)}} = \frac{(E)}{(B)} \cdot \frac{v_{(B)}}{v_{(E)}}$$

and moreover, in view of equation (1) and given that $v_{(B)} = v_f$,

$$\frac{t_{(E)}}{t_{(B)}} = \frac{TE}{BE} \cdot \frac{Z\Sigma}{BD} \qquad \text{or} \qquad \frac{t_{(E)}}{t_{(B)}} = \frac{GB}{BE} \cdot \frac{BF}{BD}$$

whence finally,

$$\frac{t_{(E)}}{t_{(B)}} = \frac{GB \cdot BF}{BE \cdot BD}$$

Thus the times taken to traverse the infinitely small parts (E) and (B) are in the ratio of the products GB · BF and BE · BD, or, as Huygens put it, in the ratio "of the ☐ GBF to the ☐ EBD," that is to say in the ratio of the "rectangle" GBF to the "rectangle" EBD.[30]

This result constitutes a first step that, from the mathematical point of view, already involved infinitely small elements, namely, the parts (E) and (B). How was it possible then to get from here to what is at issue in this demonstration: the comparison of the time it takes to travel through the entire trajectory KEZ with that for the entire trajectory AZ?

To handle this delicate problem, Huygens introduces the curve LXN defined by

$$\frac{BX}{BF} = \frac{t_{(E)}}{t_{(B)}}$$

That is,

$$\frac{BX}{BF} = \frac{BG \cdot BF}{BE \cdot BD}$$

The object of this new curve LXN is therefore to furnish a geometric representation of the ratio of the times for passing through the infinitely small parts (E) and (B).[31]

Given this, if one now considers all the infinitesimal parts (B) as equal and, additionally, as all being traversed with the same constant speed $v_{(B)} = v_f$, then all the times $t_{(B)}$ are equal, and BF is a constant. It follows that the ordinates BX express (or represent) the times $t_{(E)}$ corresponding to the times for traversing each infinitely small part of the arc KEZ.

Huygens then regarded the sum of the times $t_{(E)}$, equal to the time of the fall along KEZ, as the sum (or as being able to be represented by the sum) of all the ordinates BX ("[. . .] all BX")—that is to say, as the area under the curve LXH, namely, the area ASPRLNγHVMZA. In the same way, the constant ordinates BF express (or represent) the times $t_{(B)}$. The sum of these times, equal to the time of the fall along AZ, is considered as the sum of all the ordinates BF ("[. . .] all BF")—that is to say, as the area AKFΣZA.

As a result,

$$\frac{\text{time of fall along KEZ}}{\text{time of traversing AZ with speed } v_f} = \frac{\text{"all BX"}}{\text{"all BF"}}$$

which is also equal to

$$\frac{\text{area ASPRLNγHVMZA}}{\text{area AKFΣZA}}$$

If one now considers the motion from A to Z as uniformly accelerated, one finally obtains

$$\frac{\text{time of fall along KEZ}}{\text{time of fall along AZ}} = \frac{\text{area ASPRLNγHVMZA}}{2(\text{area AKFΣZ})}$$

or, to recall Huygens's terms:

> As the ☐ EBD is to the ☐ FBG, so is BF to BX, whence, as all the BX are to all the BF, so the time of traversing KZ is to the time of traversing AZ with the speed generated by AZ. And the time of traversing KZ is to the time of traversing AZ as the infinite space ASPRLNγHVMZA is to 2 ☐ KZ.[32]

In the following pages of the manuscript, Huygens took advantage of this essential result in such a way that, by the use of very careful arguments of the same type, he was able to arrive at his remarkable result regarding the isochronism of falls in the cycloid.[33]

The decisive aspect of Huygens's work was the daring and quite un-Euclidean mathematical approach developed there. The question remained how infinite sums of infinitely small parts or ordinates were to be interpreted on a rigorously mathematical conceptual basis. This, however, immediately brought up again all the old problems connected with manipulations of the infinite, plunging reason into the greatest uncertainties, as Galileo (speaking through Salviati) had pointed out in the *Discorsi:* "[. . .] let us remember that we are among infinites and indivisibles, the former incomprehensible to our finite understanding by reason of their largeness, and the latter by their smallness. Yet we see that human reason does not want to abstain from giddying itself about them."[34]

While Galileo, in the pages following this citation, undertook a genuine analysis of the problem of the continuum and the paradoxes of infinity, Huygens tried to avoid direct approaches that used infinitesimals and their sums in favor of elegant but much more traditional procedures, which, by contrast, were Euclidean in inspiration. He had succeeded in this aim by the end of December 1659 but was not to publish the definitive results of his work until 1673 in the *Horologium Oscillatorium.*

3. The Deductive Scheme of the Science of the Motion of Heavy Bodies

Christiaan Huygens was elected to the Royal Academy of Sciences in Paris in 1666, the year it was founded.[35] He became one of its most influential members, responsible in large part for the drafting of its research programs.[36] It was in the context of these academic duties that seven years after his election, in 1673, he published the *Horologium Oscillatorium,* without any doubt his most remarkable work, dedicating it to Louis XIV.[37]

The book is divided into five parts. With the exception of the study of the phenomenon of shocks, on which Huygens had previously expressed his views,[38] it brings together in an ordered way the whole of the research relating to the science of motion as it had been developed up to that point by the author and his contemporaries.[39]

The second part of the *Horologium* is devoted to the fall of heavy bodies and cycloidal motion. This part amounts to a recasting, in a much different mathematical style, of the investigations on the same subject that we have analyzed in earlier pages. Although in 1659 the mathematical work was truly inventive, leaping ahead of prior treatments with little care for the requirements of rigor, in the *Horologium* Huygens proposed a highly methodical formulation relying on the classical procedures of geometry and avoiding, in particular, recourse to infinite sums.

This reorganization rested in the first place on the introduction of three principles, or "Hypotheses."

The first was what is now called the principle of inertia, or the conservation of uniform rectilinear motion. The second and third together articulate the principle of composite forces and velocities, or the independence of motions:

Hypotheses

I

If there were no gravity, and if the air did not impede the motion of bodies, then any body will continue its given motion with uniform velocity in a straight line.

II

By the action of gravity, whatever its sources, it happens that bodies are moved by a motion composed both of a uniform motion in one direction or another and of a motion downward due to gravity.

III

These two motions can be considered separately, with neither being impeded by the other.

Let C be a heavy body which, starting from rest, crosses the distance CB by the force of gravity in a certain time F. And let the same body be imagined to undergo another motion by which, assuming that gravity does not exist, it crosses the straight line CD with a uniform motion in the same time F. When the force of gravity is added, the body will not move from C to D in the time F but rather to some point E vertically below D such that the distance DE equals the distance CB. And thus the uniform motion and the motion due to gravity each make a contribution, and neither impedes the other. In what follows later we will define the line in which the body moves with this composite motion when the uniform motion is neither straight up or down but in an oblique direction. But when the uniform motion CD occurs downward on the perpendicular, it is obvious that the line CD is increased by the straight line DE when the motion due to gravity is added. Likewise, when the uniform motion CD is directed upward, CD is decreased by the straight line DE, so that, for example, after the time F the body will always be found at point E. Thus, if we consider the two motions separately in each case, as we said, and if we recognize that neither motion is in any way impeded by the other, then from this we can discover the cause and the laws of acceleration of heavy falling bodies.[40]

The first "Hypothesis," the principle of inertia, could not have been formulated in its full scope by Galileo since, as Alexandre Koyré[41] and Maurice Clavelin[42] have very clearly shown, he did not succeed in conceiving of a physical body free of gravity or, correspondingly, in freeing himself from the obsession with circular motion (which we will take up in chapter 2). Though he regards this principle only as a mathematical approximation, he nonetheless gives it a working formulation, first on the third day of the *Discorsi,* then at the beginning of the fourth day, before treating the motion of projectiles:

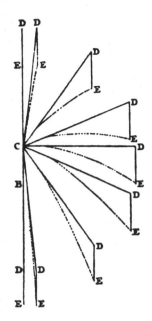

- On the third day:

> It may also be noted that whatever degree of speed is found in the moveable, this is by its nature [*suapte natura*] indelibly impressed on it when external causes of acceleration or retardation are removed, which occurs only on the horizontal plane.[43]

- On the fourth day:

> I mentally conceive of some moveable projected on a horizontal plane, all impediments being put aside. Now it is evident from what has been said elsewhere at greater length that equable motion on this plane would be perpetual if the plane were of infinite extent; but if we assume it to be ended, and [situated] on high, the moveable (which I conceive of as being endowed with heaviness), driven to the end of this plane and going on further, adds on to its previous equable and indelible motion that downward tendency which it has from its own heaviness. Thus there emerges a certain motion, compounded from equable horizontal and from naturally accelerated downward [motion], which I call "projection." We shall demonstrate some of its properties [*accidentia*] [. . .].[44]

The principle of the conservation of uniform rectilinear motion figures, by contrast, in Descartes's *Principia Philosophiae,* though the context of the physics of the plenum scarcely favored the statement of such a principle:

39. The second law of nature: all motion is in itself rectilinear; and hence any body moving in a circle always tends to move away from the center of the circle which it describes.

The second law is that every piece of matter, considered in itself, always tends to continue moving, not in any oblique path, but only in a straight line. This is true despite the fact that many particles are often forcibly deflected by the impact of other bodies; and, as I have said above, in any motion the result of all the matter moving simultaneously is a kind of circle. The reason for this second rule is the same as the reason for the first rule, namely the immutability and simplicity of the operation by which God preserves motion in matter. For he always preserves the motion in the precise form in which it is occurring at the very moment when he preserves it, without taking any account of the motion which was occurring a little while earlier.[45]

The situation was very different with regard to the principle of the components of motion. Galileo, contrary to his argument in the case of the principle of inertia, gave a very clear statement of this principle in its full scope in the opening lines of the fourth day of the *Discorsi:*

In the studies on which I now enter, I shall try to present certain leading essentials, and to establish them by firm demonstrations, bearing on a moveable when its motion is compounded from two movements; that is, when it is moved equably and is also naturally accelerated. Of this kind appear to be those which we speak of as projections [. . .].

[. . .] the moveable (which I conceive of as being endowed with heaviness), driven to the end of this plane and going on further, adds on to its previous equable and indelible motion that downward tendency which it has from its own heaviness. Thus there emerges a certain motion, compounded from equable horizontal and from naturally accelerated downward [motion], which I call "projection." We shall demonstrate some of its properties [*accidentia*], of which the first is this:

Proposition 1. Theorem 1
When a projectile is carried in motion compounded from equable horizontal and from naturally accelerated downward [motions], it describes a semiparabolic in its movement.[46]

Huygens must be credited, finally, with having perfectly grasped the role of these principles in the organization of a deductive science of the motion of heavy bodies.

On the basis of this set of "Hypotheses," completed with the working out of the implications of the third hypothesis,[47] Huygens proposed to "discover the cause and the laws of acceleration of heavy falling bodies."[48]

How, and on the basis of what mathematical treatment, was that possible? Let us follow Huygens at some length as he constructs and links together his propositions.

Proposition I

In equal times equal amounts of velocity are added to a falling body, and in equal times the distances crossed by a body falling from rest are successively increased by an equal amount.[49]

Huygens supposes that a body, departing from rest in A, falls in the course of the first interval of time as far as B and that it acquires in the course of this fall a speed such that it next traverses, by a uniform motion, a certain space BD. Now this body is also subject to the action of gravity. It is therefore animated by a "motion composed of the uniform motion by which it would have crossed BD and of a motion characteristic of falling bodies" by which "it necessarily falls" through a distance DE equal to AB. It traverses, therefore, during the second interval of time, the space AE. Since the motion acquired in B is conserved and the motion "due to gravity" in the course of the second period of time, equal to the first, is evidently the same, the motion acquired in E is twice that acquired in B. This reasoning can be repeated in the case of the third period of time, and then of the fourth, in such a way that "it is clear that whatever distances we take to be crossed successively in equal times, these distances will each increase by an amount equal to BD, and simultaneously the velocities will also be increased equally in equal times." Thus:

Proposition II

The distance crossed in a certain time by a body beginning to fall from rest is one-half the distance which it would cross in an equal time with a uniform motion whose velocity is equal to the velocity acquired at the last moment of the fall.[50]

Huygens's second proposition corresponds to the statement of the theorem of the mean degree, which is to say to the section of proposition I of the third day of Galileo's *Discorsi* relating to uniformly accelerated motion. If one considers the accompanying figure and the results obtained in the first proposition, it must now be demonstrated that the distance BD is twice the distance AB. Huygens's demonstration presents no mathematical difficulties, relying only on the classical theory of proportions:

> The distances crossed by a falling body in the first four equal times are AB, BE, EG, and GH. There is a certain proportion between these distances. Let us suppose times double these times so that, for example, for the first time we take the two times in which the distances AB and BE were crossed, and for the second time we take the two remaining times in which the distances EG and GK were crossed. It is then necessary that the distances AE and EK, which are crossed from rest in equal times, are related to each other as are the instances AB and BE, which are also crossed from rest in equal times.
>
> From this it follows that AB is to BE as AE is to EK. By conversion KE is to EA as EB or DA is to DB. Also by division DB is to BA as the excess of KE over AE is to EA. From what was proven in the prior proposition, it follows that KE is equal to twice AB plus five times BD. But EA is equal to twice AB plus BD. Thus it is clear that the mentioned excess of KE over EA is equal to four times BD. Hence DB will be to BA as four times DB is to EA. Thus EA will be four times BA. But EA is equal, as we said, to twice AB plus BD. Therefore BD is equal to twice AB. Q.E.D.[51]

Proposition III enunciates the time-squared law:

Proposition III

If two distances are crossed by a falling body in any times, each of which is measured from the beginning of the fall, these distances are related to each other as the duplicate ratio of these times, or as the squares of the times, or as the squares of the velocities acquired at the end of these times.[52]

From proposition II, previously demonstrated, Huygens immediately draws the rule of odd numbers:

> For it was shown in the preceding proposition that the distances AB, BE, EG, and GK, whatever their size, which are crossed in equal times by a falling body, increase by the same amount, which is equal to BD. Now it is clear that BD is twice AB, that BE is triple AB, that EG is five times AB, that GK is seven times AB, and that the remaining distances are successively increased according to the progression of odd numbers beginning from one, i.e., 1, 3, 5, 7, 9, etc.[53]

From this he concludes:

And since any sum of these numbers, taken consecutively, makes a square whose side equals the number of the numbers taken (for example, if the first three are added, they make nine; if four, sixteen), it follows from this that the distances crossed by a falling body, each of which is taken from the beginning of the fall, are related to each other as the duplicate ratio of the times during which the fall occurs, if indeed the times are assumed to be commensurable.[54]

He next extends this result to incommensurable times:

But it is easy to extend this demonstration to incommensurable times. Let there be two such times whose ratio is expressed by the lines AB and CD, and let the distances crossed in these times be E and F, each taken from the beginning of the fall. Now I say that the distance E is related to F as the square of AB is related to the square of CD.[55]

Huygens goes on to establish a further proposition:

Proposition IV

If a heavy body begins to move upwards with the same velocity acquired at the end of a descent, then in equal parts of time it will cross the same distances upwards as it did downwards, and it will rise to the same height from which it descended. Also in equal parts of time it will lose equal amounts of velocity.[56]

However, the reconstruction of the Galilean science of motion, consistent with the requirements of rigor enforced by Euclidean geometry, is still more evident in proposition V. This proposition is only a new demonstration of proposition II, which is to say of the "Proposition I. Theorem I" of the *Discorsi*. But, contrary to his procedure in the preceding demonstration, Huygens announces here his intention of "using Galileo's method" in the interest of being able to "write down here more accurately the demonstration which he gave in less perfect form."[57]

In so doing, Huygens constructs a demonstration that is rigorous in every respect but conceptually less rich than Galileo's. Before examining this analysis in greater detail, it will be instructive to read Huygens's splendid, but very long, text:

Proposition V

The distance crossed in a certain time by a body which begins its fall from rest is half the distance which it would cross in an equal time with a uniform motion having the velocity acquired at the last moment of the fall.

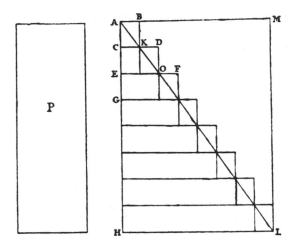

Let AH be the total time of fall. In that time the moving body crosses a distance whose quantity is designated by the plane P. Draw HL, of any length, perpendicular to AH, and let it represent the velocity acquired at the end of the fall. Next complete the rectangle AHLM, and let it represent the amount of distance which would be crossed in the time AH with the velocity HL. What must be shown is that the plane P is half of the rectangle MH, or, after drawing the diagonal AL, that the plane P equals the triangle AHL.

If the plane P is not equal to the triangle AHL, then it will be either smaller or larger. First assume that the plane P is smaller than the triangle AHL, if that be possible. Let AH be divided in a number of equal parts AC, CE, ED, etc. And on the triangle AHL let a figure be circumscribed which is composed of rectangles whose altitude equals each part of this division of AH, namely, the rectangles BC, DE, FG. Also within the same triangle let there be inscribed another figure composed of rectangles of the same altitude, namely, KE, OG, etc. Let all this be done in such a way that the excess of the former figure over the latter figure is less than the excess of the triangle AHL over the plane P. It is clear that this can be done since the entire excess of the circumscribed figure over the inscribed figure equals the lowest rectangle, which has HL for a base. Therefore the whole excess of the triangle AHL over the inscribed figure will be less than its excess over the plane P, and thus the figure inscribed in the triangle is larger than the plane P.

Further, since the straight line AH represents the entire time of fall, its equal parts AC, CE, EG, represent equal parts of that time. And since the velocities of falling bodies increase in the same proportion as the times of fall, and since HL is the velocity acquired at the end of the whole time, CK will be the velocity acquired at the end of the first part of the time, for AH is to AC as HL is to CK. Likewise EO will be the velocity acquired at the end of the second part of the time, and so forth. But it is clear that in the first time AC some distance larger than zero is crossed by the moving body. And in the second time CE the distance crossed is greater than KE since the distance KE would have been crossed in the time CE by a uniform motion with the velocity CK. For the distances crossed by a uniform motion have a ratio composed of the ratios of the times and of the velocities. And hence, since in the time AH the distance MH would have been crossed with the uniform velocity HL, it follows that in the time CE the distance KE would be crossed with the velocity CK, since the ratio of the rectangle MH to the rectangle KE is composed of the ratios of AH to CE and of HL to CK.

Now, as was said, the distance KE would be crossed in the time CE with the uniform velocity CK, but the body is moved in the time CE with an accelerated motion which already has the velocity CK at the beginning of this time. Hence it is clear that, in this accelerated motion in the time CE, a greater distance than KE is crossed. For the same reason, in the third period of time EG, a greater distance than OG is crossed since this is what would be crossed in that time EG with the uniform velocity EO. And so successively, in each part of the time AH, the body crosses a greater distance than the rectangles inscribed in the figure adjacent to each part of the time. Hence the whole distance crossed by an accelerated motion will be greater than the inscribed figure. But that distance was assumed to be equal to the plane P. Thus the inscribed figure will be smaller than the distance P, which is absurd because this was shown to be larger than the distance. Thus the plane P is not smaller than the triangle AHL. Next it will be shown that it also is not larger.

Assume that it is larger, if that be possible. Let AH be divided into equal parts, and using these parts as altitudes, let figures composed of rectangles be inscribed and circumscribed again on the triangle AHL, as before. Let this be done in such a way that the one exceeds the other by an amount less than the amount by which the plane P exceeds the triangle AHL. Hence the circumscribed figure will necessarily be less than the plane P. Now it is clear that in the first part of the time AC the distance crossed by the body is less than BC, since this would be crossed in the same time with the uniform velocity CK which the body acquires only at the end of the time AC. Likewise in the second part of the time CE the distance crossed by an accelerated motion is less than DE since this would be crossed in the same time CE with the uniform velocity EO which it acquires only at the end of the time CE. And in like fashion, in each part of the time AH, less distance is crossed by the moving body than is represented by the rectangles circumscribed on the figure adjacent to each part of the time. Hence the whole distance crossed by an accelerated motion will be less than this circumscribed figure. But that distance was assumed to be equal to the plane P. Thus the plane P will be less than the circumscribed figure, which is absurd since this figure was shown to be less than the plane P. Therefore the plane P is not larger than the triangle AHL. But it was also shown that it is not less than that triangle. Therefore it must be equal. Q.E.D.[58]

What does this long demonstration reveal? Essentially this: that Huygens, in order to lay out his argument, which is rigorous in every respect and relies on two successive reductions to the absurd, had to renounce the direct Galilean approach. Galileo's "uncertain" reasoning consisted (as we shall see in chapter 3) in directly accounting for distance covered on the basis of speed. Huygens's strategy, though it did involve the proportionality of speed to time, was feasible only to the extent that it immediately substituted distances for time. Huygens's reasoning was, in a manner of speaking, static.

The enormity of what was at stake—how mathematics was to be employed in the process of geometrization—is fully apparent here: on the one hand, we have Galileo, who sought to explain the essential processes involved in the phenomenon of the fall of heavy bodies using a new but undeveloped mathematical technique; and on the other, Huygens, who chose a traditional mode of demonstration (though earlier we have seen he was capable of taking the same mathematical risks as Galileo) to give the science of motion a rigorous formulation perfectly in accordance with the Euclidean deductive method.

This being so, Huygens hastened to add that "everything that has been demonstrated up to this point is equally true both of bodies falling and rising on inclined planes and of bodies moved on the perpendicular."[59] The way was now clear to demonstrate once again his earlier results on curvilinear trajectories and, more specifically, those relating to the isochronism of falls in the cycloid, only now using, not the fertile and heuristic—though unrigorous—methods of 1659, but rather the techniques of traditional geometry. This study occupied the remainder of the second part of the *Horologium,* that is, propositions VI to XXVI (keeping in mind that the second part included the propositions about inclined planes as well as the investigation into certain properties of the cycloid). The next to last of these propositions gives the decisive result:

Proposition XXV

On a cycloid whose axis is erected on the perpendicular and whose vertex is located at the bottom, the times of descent, in which a body arrives at the lowest point at the vertex after having departed from any point on the cycloid, are equal to each other; and these times are related to the time of a perpendicular fall through the whole axis of the cycloid with the same ratio by which the semicircumference of a circle is related to its diameter.[60]

With this proposition the theory of the motion of heavy bodies suddenly awakened to find that it was now part of Euclidean mathematics. Its principles had been discerned, and the theory had been developed step-by-step through a series of clear demonstrations. All of this had been achieved, however, at the price of forswearing all procedures that explicitly referred to the infinite. Mathematical creation had not been fully accepted. It was necessary therefore to start all over again, departing this time not from Huygen's Euclidean world but, as we shall see in chapter 3, from the direct demonstrations given by Galileo.

Two

First and Last Ratios in the Newtonian
Theory of Central Forces

1. The Construction of Circular Motion

The first decades of the seventeenth century saw the status of circular motion undergo a fundamental transformation. No longer taken as something given, it became something that had to be explained—now as an object of mathematical physics.

Since ancient times (more specifically, since the conceptual treatment given it by Aristotle), circular motion was conceived both as primary motion and as natural motion. Aristotle is very clear on this point in the *Physics:*

> It can now be shown plainly that rotation is the primary locomotion. Every locomotion, as we said before, is either rotatory or rectilinear or a compound of the two; and the two former must be prior to the last, since they are the elements of which the latter consists. Moreover rotatory locomotion is prior to rectilinear locomotion, because it is more simple and complete. For the line traversed in rectilinear motion cannot be infinite; for there is no such thing as an infinite straight line; and even if there were, it would not be traversed by anything in motion; for the impossible does not happen and it is impossible to traverse an infinite distance. On the other hand rectilinear motion on a finite straight line is composite if it turns back, i.e. two motions, while if it does not turn back it is incomplete and perishable; and in the order of nature, of definition, and of time alike the complete is prior to the incomplete and the imperishable to the perishable. Again, a motion that admits of being eternal is prior to one that does not. Now rotatory motion can be eternal; but no other motion, whether locomotion or motion of any other kind, can be so, since in all of them rest must occur, and with the occurrence of rest the motion has perished.

The result at which we have arrived, that rotatory motion is single and continuous, and rectilinear motion is not, is a reasonable one. In rectilinear motion we have a definite beginning, end and middle, which all have their place in it in such a way that there is a point from which that which is in motion will begin and a point at which it will end (for when anything is at the limits of its course, whether at the whence or at the whither, it is in a state of rest). On the other hand in circular motion there are no such definite points; for why should any one point on the line be a limit rather than any other? Any one point as much as any other is alike beginning, middle, and end, so that they are both always and never at a beginning and at an end [. . .]. Our next point gives a convertible result: on the one hand, because rotation is the measure of motions it must be the primary motion (for all things are measured by what is primary): on the other hand, because rotation is the primary motion it is the measure of all other motions. Again, rotatory motion is also the only motion that admits of being regular. In rectilinear locomotion the motion of things in leaving the beginning is not uniform with their motion in approaching the end, since the velocity of a thing always increases proportionately as it removes itself farther from its position of rest; on the other hand rotatory motion alone has by nature no beginning or end in itself but only outside.[1]

Circular motion is most perfectly embodied in nature by the uniform movement of the stars and celestial spheres; the sublunar world, by contrast, is liable to violent and perishable motions. Moreover, the natural motion of the stars, for example, proceeds by means of an internal principle, without requiring, as interrupted or accelerated motion does, the action of an external force. The motion of a cast stone and the motion of the planets are therefore intrinsically different motions.

Uniform circular motion continued to enjoy a privileged position for centuries, notwithstanding the fact that, strictly speaking, the description of planetary motion cannot be reduced to such a model. Accordingly, in order to remedy this state of affairs, ancient and Ptolemaic astronomy were led to introduce subtle combinations of uniform circular motions. This primacy of uniform circular motion was still to be found in the *De revolutionibus* of Nicholas Copernicus (1473–1543), even though by the time of its publication in 1543 the sun now stood—according to the simplest interpretation of the Copernican system—at the center of the world:

4. That the Motion of the Heavenly Bodies is Uniform, Circular, and Perpetual, or Composed of Circular Motions.

We now note that the motion of heavenly bodies is circular. Rotation is natural to a sphere and by that very act is its shape expressed. For here we deal with the simplest kind of body, wherein neither beginning nor end may be discerned nor, if it rotate ever in the same place, may the one be distinguished from the other.[2]

With the beginning of the seventeenth century the ideal supremacy of the circle and the sphere, and with it the obsession with circular motion, began gradually to fade away, undermining the special status of circular motion. Now that the earth and the planets enjoyed the same status, being subject in every case to the same laws, this motion was no longer regarded as simple and natural but as something needing to be explained; now reduced to an effect, it lost its splendid primacy, which passed to rectilinear motion.

Without entering here into a discussion of the historical events that culminated in the remarkable work of Christiaan Huygens, mention must be made, even in a cursory survey of the scientific landmarks of the first half of the seventeenth century, of certain outstanding figures.

Foremost among them is Johannes Kepler (1571–1630). In his *Epitome astronomiae Copernicae,* inspired by the magnetist philosophy of William Gilbert (1540–1603) and published sequentially between 1617 and 1621, Kepler introduced the notion of a "driving force" centered in the sun and acting upon the planets in inverse relation to distance:

> *What is it that makes the planets go round the Sun, each within the limits of its own region, if there be no solid spheres, if the globes themselves can do nothing but remain stationary, and if they cannot be moved from place to place by any soul in the absence of the solid spheres?*—[. . .] it is fairly clear that we should not have recourse either to a mind which would make the globes rotate by the dictates of reason, or as it were by order; or to a soul which would preside over this motion of revolution, and which would give the impression [of motion] on the globes through a uniform thrust of forces, as occurs in rotation about the axis. [The origin of] the motion of the primary planets round the sun is to be attributed solely and uniquely to the body of the Sun placed in the middle of the whole Universe.[3]

Galileo had also raised the problem of the nature of circular motion in the first and second days of his *Dialogo* (1632), but without managing to wholly free himself from traditional conceptions. It was Descartes who prepared the way for the modern explanation in his *Principia Philosophiae* (1644). Arguing from the observed motion of a sling with reference to the concept of "tendency to motion," devised to explain the path followed by an object once its motion has been impeded, he advanced (in paragraphs 57–59) a novel analysis of rotation. The decomposition of motion proposed by Descartes, which was not the less pertinent or suggestive for being qualitative, relied on his prior formulation of the principle of inertia, or conservation of uniform rectilinear motion:

> 57. How the same body can be said to strive to move in different directions at the same time.
>
> Often many different causes act simultaneously on the same body, and one may hinder the effect of another. So, depending on the causes we are considering, we may say that the body is tending or striving to move in different directions at the same time. For example, the stone A in the sling EA which is swung about the centre E tends to go from A to B, if we consider all the causes which go to determine its motion, since it does in fact go in this direction. But if we concentrate simply on the power of moving which is in the stone itself, we shall say that when it is at point A it tends towards C, in accordance with the law stated above (supposing, of course, that the line AC is a straight line which touches the circle at point A). For if the stone were to leave the sling at the exact moment when it arrived from L at point A,

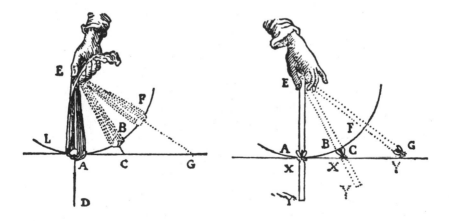

it would in fact go from A towards C, not towards B; and although the sling may prevent this outcome, it does not prevent the "striving." Finally, if we concentrate not on the stone's total power of moving but only on that part which is checked by the sling, and we distinguish this from the remaining part which produces the actual result, we shall say that when the stone is at point A, it tends to move simply towards D, or that it "strives" to move away from the centre E along the straight line EAD.[4]

Huygens's work was profoundly influenced by Cartesian thought. The outcome of his research on the law governing the fall of heavy bodies was that in 1659 he was able to give a quantitative expression for centrifugal force. His results were to be made public, however, only many years later. They initially appeared in the final pages of his *Horologium Oscillatorium* (1673); unfortunately, in these few pages Huygens limited himself to stating the thirteen theorems relating to centrifugal force, without providing demonstrations of them. The publication of these demonstrations was put off till a later date. The full treatise, drafted in 1659, was not published until 1703, eight years after Huygens's death, in the *Opuscula Posthuma* (the various texts of which had been assembled by the executors of Huygens's estate, Volder and Fullenius) under the title *De vi Centrifuga*.[5]

The statements of propositions II and III in *De vi Centrifuga* give the quantitative value of centrifugal force as a function of the speed of rotation of the moving body and of the ray of the circle described:

Proposition II

When equal moving bodies turn in the same or equal circumferences or wheels with different speeds but both with uniform motion, the centrifugal force of the faster one will be to that of the slower one in a ratio equal to that of the squares of the speeds. That is to say, if the strings by which the moving bodies are held pass from top to bottom through the center of the wheel, bearing weights by which the centifugal force of the moving bodies is held in check and exactly balanced, these weights will be to each other as the squares of the speeds.[6]

Proposition III

When two equal moving bodies move with the same speed along unequal circumferences, their centrifugal forces will be inversely proportional to the diameters, such that in the case of the smaller circumference this force will be the greater [of the two].

With Huygens, uniform circular motion was no longer simple and natural. In order to be explained, it had to be construed in such a way that the intervention of the principle of inertia was implied. This construction thus made possible the application of mathematical thought to theoretical problems that would finally culminate in Newton's *Principia*.

2. Mechanist Interlude: Centrifugal Force and Weight

In his letter of 11 October 1638 to Marin Mersenne, Descartes described the impression that Galileo's *Discorsi* had made upon him. Descartes begins by reproaching Galileo for having "built without foundation" since, on his view, Galileo "has only sought the reasons of some particular effects" without considering "the first causes of nature." [8]

Thus, in connection with weight and the fall of heavy bodies, Descartes very sternly emphasizes:

> Everything that he says about the velocity of bodies that descend in a vacuum, etc. [. . .] is constructed without foundation; for he should have previously determined what [weight] is; and if he knew the truth of it, he would know that it is nil in a vacuum. [9]

In Descartes's whirling, full universe, weight is due to the excess of motion—compared with "the parts of the earth"—of the parts of the "subtle" (or celestial) matter surrounding and driving the earth. This excess of motion tends to push the subtle matter away from the center around which it turns and thus invites other bodies to come to replace it, that is, to fall to earth. Descartes is very explicit in the eleventh chapter of *Le Monde, ou Traité de la lumière*, published for the first time posthumously in 1664 (a second edition followed in 1677) but written in the 1630s:

> But, in order to understand this more clearly, consider the earth EFGH with water 1234 and air 5678, which (as I shall tell you below) are composed simply of some of the less solid of the earth's parts and constitute a single mass with it. Then consider also the matter of the heaven, which fills not only all the space between the circles ABCD and 5678 but also all the small intervals below it among the parts of the air, the water, and the earth. And imagine that, as that heaven and this earth turn together about center T, all their parts tend to move away from it, but those of the heaven much more quickly than those of the earth,

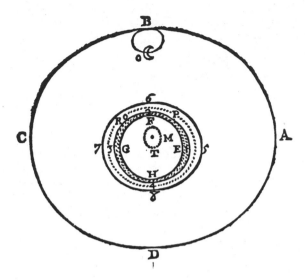

because the former are much more agitated. Or, indeed, imagine that, among the parts of the earth, those more agitated in the same direction as those of the heaven tend to move away from the center more than do the others. Thus, if the whole space beyond circle ABCD were void, i.e., were filled only with a matter that could not resist the action of other bodies nor produce any considerable effect (for it is thus that we must construe the name "void"), then all the parts of the heaven in the circle ABCD would be the first to leave it; then those of the air and of the water would follow them, and finally also those of the earth, each that much sooner as it were less attached to the rest of its mass. In the same way, a stone leaves the sling in which it is being moved as soon as one releases the cord, and the dust one throws on a top while it is turning immediately fl[ie]s off from it in all directions.

Then consider that, since there is no such space beyond circle ABCD that is void and where the parts of the heaven contained within that circle can go, unless at the same instant others completely like them enter in their place, the parts of the earth also cannot move any farther than they do from center T, unless they descend in their place just as many parts of the heaven or other terrestrial parts as are needed to fill it. Nor, in turn, can they move closer to the center unless just as many others rise in their stead. Thus they are all opposed to one another, each to those that must enter in its place in case that it should rise, and similarly to those that must enter therein in the case that it should descend, just as the two sides of a balance [scale] are opposed to one another. That is to say, just as one side of a balance can be raised or lowered only if the other side does exactly the contrary at the same instant and just as the heavier always raises the lighter, so too the stone R, for example, is so opposed to the quantity (exactly equal in size) of all above it, whose place it should occupy in the case that it were to move farther away from center T, that the air would necessarily have to descend to the extent that the stone rose. And, in the same way, it is also so opposed to another like quantity of air below it, whose place it should occupy in the case that it were to move closer to that center, that the stone must descend when this air rises.

Now, it is evident that, since this stone contains in it much more of the matter of the earth than a quantity of air of equal extent—and in recompense contains that much less of the matter of the heaven—and since also its terrestrial parts are less agitated by the matter of the heaven than those of the air, the stone should not have the force to rise above that quantity of air, but on the contrary the quantity of air should have the force to make the stone fall downward. Thus, that quantity of air is light when compared with the stone but is heavy when instead it is compared with the wholly pure matter of the heaven. And so you see that each part of terrestrial bodies is pressed toward T, not indifferently by the whole matter surrounding it, but only by a quantity of this matter exactly equal to the size of the part; that quantity, being underneath the part, can take its place in the case that the part falls. That is the reason why, among the parts of any single body designated "homogeneous" (such as among those of air or water), the lowest are not notably more pressed than the highest, and why a man down below in very deep water does not feel it weigh on his back any more than if he were swimming right on top.[10]

The weight of a body is therefore in no case a property of the body or an action of the body of the planet on the body but is a mechanical effect provoked solely by the movements of the subtle matter. Descartes thus fully satisfied the requirements of his mechanist philosophy, which stipulated that the explanation of every phenomenon must be given in terms of figure and of motion:

For I freely acknowledge that I recognize no matter in corporeal things apart from that which the geometers call quantity, and take as the object of their demonstrations, i.e. that to which every kind of division, shape and motion is applicable. Moreover, my consideration of such matter involves absolutely nothing apart from these divisions, shapes and motions; and even with regard to these, I will admit as true only what has been deduced from indubitable common notions so evidently that it is fit to be considered as a mathematical demonstration. And since all natural phenomena can be explained in this way, as will become clear in what follows, I do not think that any other principles are either admissible or desirable in physics.[11]

In 1669, Huygens, using his results on the measure of centrifugal force, tried to refine the Cartesian mechanical model of weight without modifying the style of these results. The same year, the newly founded Royal Academy of Sciences undertook a major collective investigation into "the causes of weight."[12] The first of a number of papers on the subject was read at the regular Wednesday session of the Academy on 7 August 1669 by Gilles Personne de Roberval (1602–1675). This paper opens with a brief presentation of the three main rival theses of the day, which were to fuel the debates of the following weeks and months.

Roberval begins by defining the weight of a body as "that which causes this body to descend to a center by nature alone and without artifice."[13] Given this, "one may consider a terrestrial weight, a lunar [weight], a solar [weight], a Jovian [weight] etc." Each planet therefore possesses its own peculiar weight, a fact that needs to be explained:

> It is not necessary to attribute a particular virtue to this center, which is only a point; but it suffices to understand that all the parts of the body are inclined to unite together to make a single body; for from that will result a center of gravity toward which all the parts will be directed, with more or less force, according to their specific nature: and weight consists in just this force.[14]

He next distinguishes between three explanatory models, or "opinions":

- "[. . .] weight resides in the heavy body alone."[15]
- Weight is "common and reciprocal between this heavy body and that [body] towards which it tends to move."[16]
- Weight is "produced by the effort of a third [body] which pushes the heavy body."[17]

Roberval, however, considering research into principles and causes to be of secondary importance, adopts a different position:

> In conclusion I will always do my utmost to imitate Archimedes, who with regard to weight took as a principle, or postulate, the fact, constant and anchored in [the experience of] all past centuries up until the present, that there are heavy bodies which have the conditions he speaks of at the beginning of his treatise on the subject,[18] and on this foundation I will establish

my arguments for Mechanics as he did, without troubling to
know the fundamental principles and causes of weight [but] limiting myself to following the truth, if one day it should wish to
reveal itself clearly and distinctly to my mind.[19]

Let us return now to the different "opinions" defended by Roberval's fellow academicians.

The first "opinion" upheld the idea that there is "in the heavy
body a quality that carries it downward,"[20] or, as Jacques Buot
(?–c. 1673) put it, "a natural and absolute virtue [. . .] by which it
descends of itself toward the earth."[21] But this opinion, according
to Buot, "finds no defenders in a celebrated company that is not
satisfied with the words [']natural virtue['], which is [said to be]
responsible for the descent of bodies, for it is the cause of this descent and the nature of this virtue that are being investigated."[22]

The second "opinion" supports the idea that weight is "an at-
tractive and mutual quality among all the parts of an entire body
for uniting themselves together as far as they may."[23] This "opinion" was defended by Bernard Frénicle de Bessy (c. 1600–1675) in
a paper delivered at the Academy's session of 14 August 1669. It
relied on a number of experiments making reference, in particular,
to the phenomenon of capillarity:

> All these experiments clearly show that there is no ground for
> doubt about this attractive virtue, or of this desire for union,
> which obliges bodies to remain attached to each other.[24]

And as a result:

> [. . .] I see nothing besides the inclination that one perceives in
> bodies to unite themselves with each other, which I have named
> attraction, that might be [the] cause of weight.
> Now if it is the mutual attraction of bodies that makes up
> weight: it follows that a particular body falling from the surface
> of the earth toward the center will travel more quickly at the be-
> ginning of its motion than if it begins to fall from a place nearer
> the center. The reason is that the inclination or virtue that bodies
> have of attracting one other, being in all the parts of the Earth,
> each one of them, acts on this body; and thus the parts of the
> earth that are near the surface and farthest from the center act
> on this body as well [and] must retard the movement that the
> rest of the earth impresses upon it; but it would be necessary [to
> descend to] a very great depth to be able to remark any difference
> in [this motion]; and the deepest gold or silver mines are not deep
> enough to provide any sensible [indication].

As all things have their own particular virtue, that which the Earth has of attracting bodies cannot be infinite, since the earth is a finite agent, and all that is finite is limited in its forces.

From this principle it follows that a terrestrial body may be thrown [up in the air] so high that it will no longer fall back to earth; otherwise the virtue of attraction would be infinite: the small bodies of which I have spoken above and which attract each other do not much extend this virtue, not more than a distance of 6 or 7 lines, because they are very small, [though] the sides of the vessel in which they are shaken attract somewhat further. [So] that if one wished to make the size of these small bodies proportional to the distance over which they attract [in relation] to the size of the earth and the distance over which it can attract; that distance would be very great and such that we could have no experience of it.[25]

Frénicle's opinion was seconded some months later by Edme Mariotte (c. 1620–1684) in a paper presented at the session of 13 November 1669:

Of all the opinions that have been proposed until now regarding the causes of weight, I find none more probable than that of M. Frenicle who has shown that the weight of terrestrial bodies is an effort of motion of these bodies toward the center of the Earth.[26]

A bit further on Mariotte adds:

One can conceive by this motion that of heavy bodies toward the center, that is to say, that they have a disposition or a virtue to move toward other bodies that is natural and adherent to them, and that does not die out though one holds them at rest, forcibly, at least when they are not too distant from other bodies[:] this motion we call [a] motion of junction or of aggregation, many other examples of which are seen in nature, [as, for example,] if a bead of quicksilver is drawn near to another until it touches, the other neighboring parts first of all conceive & [then] take a motion toward one another until they form a single ball[;] water and air do the same, but if one brings a bead of quicksilver close to one of Water, they touch without either one moving toward the other, [so] that if it is said that the pores of the water and of the quicksilver are not disposed [in] the same [way] & that their respective parts cannot enter into the pores of the other, the answer is that it does not suffice for an ankle to be next to a hole proportionate to its size and shape for it to go in, but that it is necessary for it to be pushed into [the hole]; one sees this respective motion also in lodestone and iron, the one [moving] toward the other when they are at a certain distance, and similarly with the steel and iron in hot iron and molten tin, for if one puts glass or wood in the molten tin it does not at all attach itself to it as [it does] to hot iron. Here then is a hypothesis that may be established, [namely,] that the earth, the bodies of the planets, such as the moon, Jupiter &c. are spherical bodies & that they would not be [spherical] if their parts did not have this motion of junction, for in turning on their axes the external parts moving along the tangents would diverge and disperse in the air; therefore it has been necessary for them to have [their own] respective motion, and the more solid and dense bodies have it more strongly than the subtle [ones] such as air &c.[27]

As for the third opinion, it introduced, in order to explain weight, "some very subtle body that moves by a very fast motion and that insinuates itself easily among the parts of other more sizable bodies, such that in pressing upon them, it pushes them downward or upward: and by this means they produce weight or lightness."[28] This "opinion" was, of course, the one upheld by Descartes and his followers. It was defended before the Academy by Buot in a paper read at the 21 August 1669 session, and also by Christiaan Huygens in a paper given the following week.[29]

Huygens immediately situated his argument within the Cartesian mechanist framework with its well-defined criteria of intelligibility:

> To seek an intelligible cause of weight it is necessary to see how it can come about, supposing in nature only bodies made of [the] same matter, in which one considers no quality, nor inclination to approach each other, but only different sizes, shapes and motions; how, I say, it can come about nonetheless that these bodies directly tend toward a common center, and hold together [being] assembled around it, which is the principal and most ordinary phenomenon of what we call Weight.[30]

For Huygens it was now a question of finding an explanation of the phenomenon of weight within this framework and this framework alone:

> The simplicity of the principles that I have just specified does not leave much choice in this research, for one judges right away that there is no appearance of attributing [either] to the shape or to the smallness of the particles any effect similar to that of weight, which being a tendency or inclination to motion must truly be produced by a motion: Such that it remains only to seek

in what manner [such a motion] can [act] and [upon] which
bodies.

   We see two sorts of motion in the world, the straight and the
circular[,] and [while] we know nothing about the nature of the
first and the laws that govern bodies in [respect of] the communi-
cation of their motions, when they come into contact, insofar as
one considers only straight motion and the reflections that arise
from it among the parts of matter one finds nothing that deter-
mines them toward a center. It must be therefore that one comes
necessarily [to consider] the properties of circular motion, and to
see that one of these may serve us in this [purpose].[31]

Huygens's explanation repeated for the most part that of Descartes, while strengthening it with results he had recently obtained
in connection with the measure of centrifugal force:

   To arrive therefore at a possible cause of weight, I shall suppose that in the spherical space which comprehends the earth &
the bodies that are around it to a great extent there is a fluid
matter which consists of very small parts and which is variously
agitated in all directions with much rapidity, [this] matter not
being able to escape this space that is surrounded by other bodies, [and so] I say that its motion must become in part circular
around the center, not so much that it comes to turn all in [the]
same direction, but such that the majority of its different motions
occur in spherical surfaces around the said center space, which
on that account becomes the center of the earth as well.

   The reason for this circular motion is that the matter contained in any space moves more easily in this manner than by
straight motions [that are] contrary to each other, which in being
reflected because matter cannot escape the space that encloses it
come necessarily to be changed into circular [ones].[32]

And, as a consequence:

   It is not difficult now to explain how weight is produced by
this motion, for if among the fluid matter that turns in space as
we have supposed are found parts much larger than those that
compose it, or bodies made up of a mass of small parts hung
together, and these bodies do not follow the rapid motion of the
said matter, they will necessarily push toward the center of the
motion and form there the terrestrial globe, if there are enough
of them for that, [it being] supposed that the earth was not yet
[created]. And the reason is the same as the one which in the
experiment described above [accounts for] the fact that the saw-
dust accumulates in the center of the vessel. It is in this therefore
that the weight of bodies consists, which one may say is the ac-

tion of the fluid matter which turns in a circle around the center of the earth in all directions, by which [the earth] tends to move away and to push in its place bodies that do not follow this motion.[33]

Huygens, after having criticized and refined the Cartesian model, goes on to discuss various quantitative studies that make it possible to determine, for example, the velocity of the subtle matter in the neighborhood of the earth, from which he then concludes:

> Besides, the great velocity of the matter not only is not contrary to reason but [actually] helps to explain further other phenomena of weight since by [means of] it one readily conceives how falling bodies always accelerate their motion even though they have already acquired [a] very great [velocity]. For that of the matter which produces weight greatly surpasses the velocity of a cannonball, for example [one] which falls back [to earth] after having been shot [into] the air perpendicularly [to the ground]; this ball until the end of its fall is almost always subject to the same pressure of this matter and consequently its velocity is continually augmented by it. So [too] if it had only little motion, the ball, after having acquired [just this] much of it, would no longer accelerate its fall, because otherwise it would be obliged to push [away] the fluid matter in order to take its place, [which would require] more velocity than it could [generate] by its own motion [alone].[34]

Whence finally:

> Lastly, the same velocity of this matter, joined in the manner we have described [so] that it acts upon the bodies that it renders heavy, shows the correctness of the principle that Galileo took to demonstrate the proportion of the acceleration of falling bodies, which is that their velocity increases equally over equal times. For these bodies being pushed successively by the parts of the neighboring matter, which tries to occupy their place and whose motion is always infinitely faster than that which [these bodies] may have acquired by the [kind of] falls with which we are familiar from experience, the result is that the action of the matter that presses upon them can be considered [as] always [being] as strong as when it finds them at rest, from which one then concludes rather easily [that] the increase of velocities [is] proportionate to that of the times.
> Having therefore demonstrated that our hypothesis contains nothing [that is] impossible and that by [means of] it one can explain all the phenomena of weight; to wit, why terrestrial bod-

ies tend toward the center [of the earth], why the action of gravity cannot be prevented by the interposition of any body [among] those that we know, why the parts within each body all contribute to its weight, and why finally heavy bodies in [the course of] falling add continually to their speed and this in proportion to the time of their descent[;] [thus] there is nothing that prevents [this hypothesis] from being true as long as one does not find other phenomena in nature that may be contrary to it.[35]

It was in the context of such views on the Continent that in 1687, in London, Newton published the *Principia,* thereby unifying the system of the world.

3. The Deductive Scheme of Newton's *Principia*

Newton's *Philosophiae naturalis principia mathematica,* a difficult work originally written in Latin and first published when he was forty-four years of age, stands today as unquestionably one of the most important achievements in the history of scientific thought.[36]

Its importance derives from the introduction of the law of universal gravitation, which made it possible to synthesize all the work done on gravitation since Galileo. According to this law, bodies are attracted with a force proportional to the product of their mass and inversely proportional to the square of the distance between them. In bringing under this single law both celestial phenomena (planetary movement) and terrestrial phenomena (the fall of bodies to earth), Newton unified physics. Henceforth the same principles, the same laws, applied equally to earth as to the heavens. The Aristotelian hierarchical cosmos was destroyed once and for all.

Beyond its extreme conceptual innovation, however, Newton's work was characterized by an unprecedented requirement of mathematicity. This reflected a desire, on the one hand, to clearly state the principles that govern theoretical advance and, on the other hand, to develop the mathematics capable of making theoretical advance possible.

3.1. The Deductive Ordering of the Principia

The *Principia* is divided into three parts or books. The first develops, from a strictly mathematical point of view, the whole set of questions pertaining to the science of motion independently of the resistance exercised by media. The results are then employed in the third book to resolve astronomical and physical questions (having to do with the motion of the planets and of the moon, the shape [or "figure"] of the earth, the theory of the tides, and so forth).

The second book is chiefly devoted to the motions of bodies in resistant media. There Newton treats the motion of projectiles in media whose resistance varies with their velocity (section I) or with the square of the velocity (section II) or with the linear combination of the two (section III).[37] He also poses the problems of the shape of the solid of least resistance (section VII) and of the theoretical justification for Torricelli's law for the velocity of fluid flowing out of a container (section VII).[38] These studies led Newton in the ninth and final section of this book to develop a vigorous critique of the Cartesian hypothesis of vortices. The style of this critique made it possible to perfectly appreciate the opposition between his physicomathematical deductive system of the world and the Cartesian geometric cosmology.

In the Cartesian system, the planets revolve around the sun, impelled by their own heaven:

> 26. That the earth is at rest in its heaven which nonetheless carries it along.
>
> Fourth, since we see that the Earth is not supported by columns or suspended in the air by means of cables but is surrounded on all sides by a very fluid heaven, let us assume that it is at rest and has no innate tendency to motion, since we see no such propensity in it. However, we must not at the same time assume that this prevents it from being carried along by [the current of] that heaven or from following the motion of the heaven without however moving itself: in the same way as a vessel, which is neither driven by wind or by oars, nor restrained by anchors, remains at rest in the middle of the ocean; although it may perhaps be imperceptibly carried along by [the ebb and flow of] this great mass of water.
>
> 27. That the same is to be believed of all the Planets.
>
> And just as the other planets resemble the Earth in being opaque and reflecting the rays of the Sun, we have reason to believe that they also resemble it in remaining at rest, each in its own part of the heaven, and that the variation we observe in their position results solely from the motion of the matter of the heaven which contains them.[39]

In this system it is readily understood that the stars farthest away revolve more slowly than the nearest ones, but the system does not specify the exact time of the revolution of each planet, how this time depends on the distance between the planet and the sun, or how the speed varies with the distance.

A satisfactory answer to these questions could not restrict itself to the Cartesian description, which situated itself at a merely qualitative level, giving only the general appearance of motions. It was necessary to express in precise terms the constitutive elements of the mechanism of vortices, in order that a conceptual framework could be constructed leading to a mathematical formulation that would permit the development of a predictive calculus. This was the task to which Newton devoted himself, with real success, in the ninth section of the second book.

If the three books whose contents we have just more or less cursorily described form the main part of Newton's work, note must nonetheless be taken of the two preliminary sections with which the *Principia* opens, placed before book I, at the entrance, as it were, of the deductive edifice proper. These two sections are entitled "Definitions" and "Axioms, or Laws of Motion," respectively.

Of particular interest is the definition, under the first rubric, of the quantity of matter:

Definition I
The quantity of matter is the measure of the same, arising from its density and bulk conjointly.[40]

And of the quantity of motion:

Definition II
The quantity of motion is the measure of the same, arising from the velocity and quantity of matter conjointly.[41]

Of impressed force *(vis impressa):*

Definition IV
An impressed force is an action exerted upon a body, in order to change its state, either of rest, or of uniform motion in a right line.[42]

And of centripetal force *(vis centripeta):*

Definition V
A centripetal force is that by which bodies are drawn or impelled, or any way tend, towards a point as to a centre.[43]

Before considering these definitions further, it is important to note two things. On the one hand, Newton, unlike Galileo (see chapter 3, below), does not give a definition of velocity, although he employs the concept; and therefore, strictly speaking, kinematics in the modern sense of the term is not to be found in Newton. On the other hand, the definition of centripetal force is not related to a particular mode of action, since it "pulls," "pushes," and "tends."

This first set of definitions (there are eight in all) is supplemented by a very important scholium that introduces, in particular, the celebrated definitions of absolute space and absolute time:

> I. Absolute, true, and mathematical time, of itself, and from its own nature, flows equably without relation to anything external, and by another name is called duration; relative, apparent, and common time, is some sensible and external (whether accurate or unequable) measure of duration by the means of motion, which is commonly used instead of true time; such as an hour, a day, a month, a year.
>
> II. Absolute space, in its own nature, without relation to anything external, remains always similar and immovable. Relative space is some movable dimension or measure of the absolute spaces; which our senses determine by its position to bodies; and which is commonly taken for immovable space; such is the dimension of a subterraneous, an aerial, or celestial space, determined by its position in respect to the earth. Absolute and relative space are the same in figure and magnitude; but they do not remain always numerically the same. For if the earth, for instance, moves, a space of our air, which relatively and in respect of the earth always remains the same, will at one time be one part of the absolute space into which the air passes; at another time it will be another part of the same, and so, absolutely understood, it will be continually changed.[44]

This first section is followed by another, entitled "Axioms, or Laws of Motion." There Newton joins together for the first time the three great laws of mechanics in a form sometimes quite close to the one in which we know them today.

The first expresses the principle of inertia, or the principle of the conservation of uniform and rectilinear motion:

Law I
Every body continues in its state of rest, or of uniform motion in a right line, unless it is compelled to change that state by forces impressed upon it.[45]

The second law stipulates the following:

Law II

The change of motion is proportional to the motive force impressed; and is made in the direction of the right line in which
that force is impressed.[46]

Obviously this law is not to be confused with the one expressed
in differential terms that we know today as "Newton's law." In particular, Newton speaks here of the "change of motion" without
any qualification regarding the time during which this change takes
place. If one were to insist on writing down this law in modern
terms, the closest expression would no doubt be $F = \Delta(mv)$, where
F is the motive force impressed, m the mass, and v the velocity,
given that $\Delta(mv)$ represents the "change of motion." In this sense
it may be said that an impressed motive force is not a force, in the
modern sense of the term, but an impulse, or impetus.

The third law asserts the equivalence of action and reaction:

Law III

To every action there is always opposed an equal reaction: or,
the mutual actions of two bodies upon each other are always
equal, and directed to contrary parts.[47]

This third law, which did not figure in the preliminary drafts of
1685 for the *Principia*,[48] allowed Newton in book 3 to formulate
the law of universal gravitation in its full scope.

It is on the basis of these definitions and laws that the propositions of book 1 are demonstrated, and thus that the motion of
bodies subject to the action of central forces is given mathematical
existence. To this end Newton applies the mathematical methods
of classical Euclidean geometry, enriched, however, on the one
hand, by numerous results relating to the study of conics (sections
IV and V, chiefly, of book 1) and, on the other hand, by the
"method of first and last ratios" presented in the first section of
the first book. This section is composed of eleven lemmas and
two scholia.

In the scholium that closes the opening section, Newton explains
that this method has been preferred to that of "the ancient geometers," in order "to avoid the tediousness of deducing involved demonstrations *ad absurdum*," but also to that of "indivisibles," which,
while having the advantage of allowing shorter demonstrations,

"seems somewhat harsh" and "ungeometrical." The method of first and last ratios therefore permitted Newton not only to avoid very long deductions such as those we met with in Huygens's *Horologium* in connection with the fall of heavy bodies but also to avoid recourse to "ungeometrical" procedures involving infinite summations and other difficulties associated with the continuum.[49]

The meaning of the method finally retained is defined thus:

> [. . .] I chose rather to reduce the demonstrations of the following Propositions to the first and last sums and ratios of nascent and evanescent quantities, that is, to the limits of those sums and ratios, and so to premise, as short as I could, the demonstrations of those limits. For hereby the same thing is performed as by the method of indivisibles; and now those principles being demonstrated, we may use them with greater safety.[50]

Lazare Carnot (1753–1823), in his *Réflexions sur la métaphysique du calcul infinitésimal* (1797), admirably summarized the spirit of the Newtonian method:

> On the method of Prime and Ultimate Ratios, or of Limits
>
> 73. The method of prime and ultimate ratios, or of limits, likewise derives its origin from the method of exhaustion; and is only, properly speaking, a development and simplification of this latter. It is to Newton that we owe this useful improvement of it, and in his book of the Principia must we look for information respecting it: our end will be gained by giving a slight sketch of it.
>
> When any two quantities are supposed to approach continually to each other, so that their ratio or quotient differs less and less gradually, and as little as we please from unity, these two quantities are said to have their ultimate ratio a ratio of equality.
>
> Generally, when we suppose that different quantities mutually and in equal portions of time approach other quantities which are considered as fixed, until they all at the same time differ as little as we please, the ratios existing between these fixed quantities are the ultimate ratios of those which are supposed to approach respectively and in equal proportions of time; and these quantities themselves are called limits or ultimate ratios of those which thus approach one another.
>
> These ultimate values and ratios are also called prime values and ratios of the quantities to which they refer, according as we consider the variables approaching to, or retiring from, the quantities which are considered fixed, and serve for their limit.[51]

Newton's statement of the first lemma is particularly illuminating:

Lemma I

Quantities, and the ratios of quantities, which in any finite
time converge to equality, and before the end of that time approach nearer to each other than by any given difference, become
ultimately equal.[52]

One of the most important results of this first section is given by
lemma VII, which states the equality, at the limit, of the arc, chord,
and tangent for every curve, as Newton puts it, of "continued curvature":

Lemma VII

The same things being supposed, I say that the ultimate ratio
of the arc, chord, and tangent, any one to any other, is the ratio
of equality.[53]

The demonstration that follows provides a supreme example not
only of the style of Newton's method but also of the care he took
to ensure the rigor of the reasoning employed:

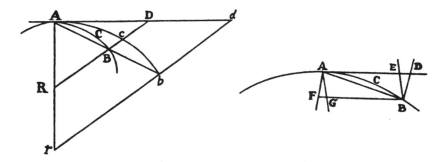

Assume, Newton says, an arc ACB "given in position" and "subtended by its chord AB" such that "in any point A, in the middle
of the continued curvature, [it] is touched by a right line AD." Considering now that the lines AB and AD are "produced to the remote
points *b* and *d;* and parallel to the secant BD draw *bd;* and let the
arc A*cb* be always similar to the arc ACB." Now that all the necessary elements are at hand—namely, the mobile elements and the
finite elements (or, in the case of magnitudes, finite magnitudes)—
the Newtonian procedure can be applied. Supposing (in accordance with lemma VI) that "the points A and B approach one another and meet, [. . .] the angle BAD [. . .] ultimately will vanish."[54]
Newton goes on to say, "[. . .] and therefore the right lines A*b*, A*d*

(which are always finite), and the intermediate arc A*cb,* will coincide, and become equal among themselves. Wherefore, the right lines AB, AD, and the intermediate arc ACB (which are always proportional to the former), will vanish, and ultimately acquire the ratio of equality. Q.E.D."[55]

The essential thing to note about this method, which was to serve as the mathematical justification for the deductive organization of the *Principia,* is that it enabled Newton to avoid long and tiresomely rigorous demonstrations in the style of the ancients, as well as the traps posed by indivisibles. By means of this lucid and rational approach, Newton was able right away to situate his undertaking within the framework of a mathematical physics that was both fertile and conscious of the imperatives of mathematical rigor. All of this is not to say that the deductive edifice that Newton was about to erect would be either without flaws or lacking in liberty with regard to the method of first and last ratios, but that the requirement of total mathematicity in the domain of mechanics had fundamentally established itself as the context within which the work of the mathematical physicist would have to be carried out.

3.2. The Theory of Central Forces

3.2.1. The Law of Areas

The study of the motion of bodies subject to central forces opens, in section II of book 1 of the *Principia,* with two propositions, the second of which is the converse of the first, establishing the characteristic property of this type of movement: centrally accelerated motions are planar motions and the area described by what we would now call the radius vector is proportional to time. It needs to be kept in mind, by the way, that the demonstrations in the *Principia,* as in other contemporary writings, relied on figures as the very basis for the arguments made, as the key to understanding the problems that gave rise to them, and not as simple representational diagrams used to illustrate the reasoning employed.

The first proposition of book 1 stipulates the following:

Proposition 1. Theorem 1
The areas which revolving bodies describe by radii drawn to an immovable centre of force do lie in the same immovable planes, and are proportional to the times in which they are described.[56]

Newton does not specify the type of curve described by the body nor whether this particular curve must be closed. (The case of the hyperbola is considered in proposition XII of section III.)

Since the body describes a curve, it follows from law I that a force must continually be impressed upon it. Newton supposed therefore that the force acting in this way is centripetal and that its center is an unmovable and mathematical point. Considering the case where the point in question is a physical, rather than a mathematical, point, Newton was led to solve another problem, involving two bodies, which required invoking the third law of motion.[57]

In this first proposition, it is necessary therefore to demonstrate first that the curvilinear trajectory is planar and next that the "radius vectors" describe areas proportional to the times. The difficulty of the demonstration resides chiefly in the mathematical treatment of the supposedly continuous action of the centripetal force. How is one to account for a continuous force when it is conceptualized by means of the fundamentally discontinuous model of percussions, or shocks?

> For suppose the time to be divided into equal parts, and in the first part of that time let the body by its innate force describe the right line AB. In the second part of that time, the same would (by Law I), if not hindered, proceed directly along to *c*, along the line B*c* equal to AB; so that by the radii AS, BS, *c*S, drawn to the center, the equal areas ASB, BS*c*, would be described.[58]

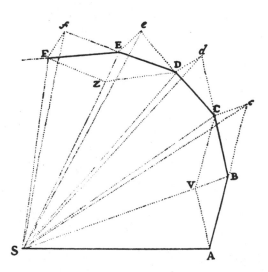

The equality of the areas of the triangles ASB and BS*c* follows from the fact that these triangles have equal bases (AB = B*c*) and the same height (the perpendicular issuing from S and falling upon AB*c*). Instead of pursuing its rectilinear trajectory from B to *c*, however, the body proceeds (in accordance with law I) toward C:

> But when the body is arrived at B, suppose that a centripetal force acts at once with a great impulse [*impulsu unico sed magno*], and, turning aside the body from the right line B*c*, compels it afterwards to continue its motion along the right line BC.[59]

Time being divided into equal parts, the force acts instantaneously at the beginning of each of these parts. This mode of action by a force is not an isolated example in the *Principia*: quite to the contrary, one encounters it again in proposition II of book 2, for example, where it allows resistance in the motion of projectiles to be conceptualized:[60]

> Case I. Let the time be divided into equal intervals; and if at the very beginning of each interval [*ipsis particularum initiis*] we suppose the resistance to act with one single impulse [*impulsu unico*] which is as the velocity, the decrement of the velocity in each of the intervals of time will be as the same velocity.[61]

Moreover, the mode of instantaneous action implied by the underlying model of shocks refers directly to law II, the statement of which contains no explicit mention of time. Let us return to proposition I:

> Draw *c*C parallel to BS, meeting BC in C; and at the end of the second part of the time, the body (by Cor. I of the Laws) will be found in C, in the same plane with the triangle ASB.[62]

The coplanarity of the motion results from the fact that *c*C is parallel to BS and that C lies in the plane defined by ASB. The motion due to central force is a planar motion:

> Join SC, and because SB and C*c* are parallel, the triangle SBC will be equal to the triangle SB*c*, and therefore also to the triangle SAB. By the like argument, if the centripetal force acts successively in C, D, E, &c., and makes the body, in each single particle of time, to describe the right lines CD, DE, EF, &c., they will all lie in the same plane [. . .].[63]

This first point established, Newton goes on to the second one, concerning the expression of the law of areas:

[. . .] and the triangle SCD will be equal to the triangle SBC, and
SDE to SCD, and SEF to SDE. And therefore, in equal times,
equal areas are described in one immovable plane: and, by composition [*componendo*], any sums SADS, SAFS, of those areas,
are to each other as the times in which they are described.[64]

The equality of the areas of the different triangles is entailed, as
before, by the fact that they have the same base and the same
height. At this stage of the rational construction, the continuity of
the action of the force and, correspondingly, the generation of the
curve are far from being established. In fact, what is generated by
this succession of actions, operating at the beginning of each small
part of time, is a polygon having the vertices A, B, C, D, and so
on. How is it possible to pass from this polygonal figure to a curvilinear figure generated by the supposedly continuous action of a
centripetal force?

Now let the number of those triangles be augmented, and their
breadth diminished *in infinitum*; and (by Cor. IV., Lem. III) their
ultimate perimeter ADF [*ultima perimeter*] will be a curved line:
and therefore the centripetal force, by which the body is continually drawn back from the tangent of this curve, will act continually [*indefiniter*]; and any described areas SADS, SAFS, which are
always proportional to the times of description, will, in this case
also, be proportional to those times. Q.E.D.[65]

The proportionality of the areas to the times being "always" realized, this proportionality is preserved at the limit and, as a consequence, a motion due to central force necessarily describes equal
areas, in the same plane, and in equal times. Neither the "necessity" of this result, though it was clearly shown by Newton, nor its
full rational import was always grasped by his contemporaries,
however.

It was by appeal to corollary 4 of lemma III of the first section,
relating to the convergence of series of inscribed figures,[66] that
Newton introduced the "continued" curvature of the trajectory
and, from that, concluded that the action of the centripetal force
was uninterrupted, or "continual." It was not readily apparent,
however, how the relation between the instances of that "great impulse" and the "continually" acting centripetal force was to be rigorously formulated. The task facing mathematical physics was far
from being achieved: strictly speaking, it lacked an analysis of the
continuum freed from geometrical intuitions.

Proposition I is supplemented by six corollaries. For our purposes here it will be useful to give the contents of corollaries 2–4 in view of their essential role in the demonstration of proposition VI of this section, which supplies the general expression of force in the case of motions due to a central force:

> Cor. II. If the chords AB, BC of two arcs, successively described in equal times by the same body, in spaces void of resistance, are completed into a parallelogram ABCV, and the diagonal BV of this parallelogram, in the position which it ultimately acquires when those arcs are diminished *in infinitum,* is produced both ways, it will pass through the centre of force.[67]

This corollary simply expresses a geometrical consequence of the force's mode of action, emphasizing that BV remains always—"ultimately"—directed toward S.

> Cor. III. If the chords AB, BC, and DE, EF, of arcs described in equal times, in spaces void of resistance, are completed into the parallelograms ABCV, DEFZ, the forces in B and E are one to the other in the ultimate ratio of the diagonals BV, EZ, when those arcs are diminished *in infinitum.* For the motions BC and EF of the body (by Cor. I of the Laws) are compounded of the motions B*c*, BV, and E*f*, EZ; but BV and EZ, which are equal to C*c* and F*f*, in the demonstration of this Proposition, were generated by the impulses of the centripetal force in B and E, and are therefore proportional to those impulses.[68]

The second corollary, with the help of the diagonals, made it possible to study the direction of forces in the course of passing to a limit; the object of this further corollary was to determine their magnitude, that is, the magnitude acquired by the forces in the course of the same passage to a limit. This result presents no difficulties, at least if the analysis of the passage to a limit is not pushed too far. Newton, in fact, considering that in equal times the motions along BC and EF are composed (in accordance with the first corollary of the laws)[69] of the inertial movement along B*c* on the one hand and E*f* on the other, and of the movements along C*c* and F*f*, concluded that since B*c*CV and E*f*FZ are parallelograms, BV and EZ, being equal to *c*C and *f*F, are to each other as the impulses in B and E.

> Cor. IV. The forces by which bodies, in spaces void of resistance, are drawn back from rectilinear motions, and turned into curvilinear orbits, are to each other as the versed sines of arcs

described in equal times; which versed sines tend to the centre of force, and bisect the chords when those arcs are diminished to infinity. For such versed sines are the halves of the diagonals mentioned in Cor. III.[70]

The diagonals introduced in corollary III are replaced here by their halves, called "versed sines."

3.2.2. The General Expression of Central Force

It is in proposition VI of the second section of book 1 that Newton gives the general expression of central forces:

Proposition VI. Theorem V

In a space void of resistance, if a body revolves in any orbit about an immovable centre, and in the least time describes any arc just then nascent; and the versed sine of that arc is supposed to be drawn bisecting the chord, and produced passing through the centre of force: the centripetal force in the middle of the arc will be directly as the versed sine and inversely as the square of the time.[71]

The demonstration that follows draws upon two results previously obtained: the fourth corollary of proposition I, which we have already mentioned, and the second and third corollaries of lemma XI of section I.

Lemma XI

The evanescent subtense of the angle of contact, in all curves which at the point of contact have a finite curvature, is ultimately as the square of the subtense of the coterminous arc.[72]

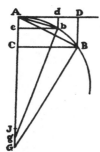

The demonstration of this lemma, which relies mainly on the existence of an osculatory circle, leads Newton to say that the last ratios from AB^2 to Ab^2 (the subtending chords, or subtenses, of the arc) and from BD to bd (the subtenses of the angle of contact) ultimately "are the same." In more modern terms, this result amounts to saying that if the chord is taken as the principal infinitely small element, then BD is an infinitely small second-order element.

It follows then:

> Cor. I. Therefore since the tangents AD, Ad, the arcs AB, Ab,
> and their sines, BC, bc, become ultimately equal to the chords
> AB, Ab, their squares will ultimately become as the subtenses
> BD, bd.[73]

This corollary means only that, in the statement of proposition
I, the chord (or the subtense of the arc) may be replaced by the arc
$\overset{\frown}{AB}$ or by the tangent AD or by the sine BC. In the fifth corollary,
Newton was to add, as if it were already clear, that the curve with
finite curvature is ultimately a parabola with vertex A.

> Cor. II. Their squares are also ultimately as the versed sines of
> the arcs, bisecting the chords, and converging to a given point.
> For those versed sines are as the subtenses BD, bd.[74]

Note that reference is not made here to the versed sines of the
arcs $\overset{\frown}{AB}$ and $\overset{\frown}{Ab}$ but to those in A of double arcs, as the first corollary of proposition VI will confirm.[75]

The third corollary gives this result a kinematic interpretation:

> Cor. III. And therefore the versed sine is as the square of the
> time in which the body will describe the arc with a given ve-
> locity.[76]

In fact, when an arc is described by a body with a given constant
velocity, the length of this arc is proportional to the duration of the
passage along it, and the versed sine is ultimately as the square of
the arc (by corollaries 1 and 2)—and so, ultimately, as the square
of the time.

This preparatory work made possible, at least in its main part,
the demonstration of proposition VI:[77]

> For the versed sine in a given time is as the force (by Cor. IV,
> Prop. I); and augmenting the time in any ratio, because the arc
> will be augmented in the same ratio, the versed sine will be aug-
> mented in the square of that ratio (by Cor. II and III, Lem. XI),
> and therefore is as the force and the square of the time. Divide
> both sides by the square of the time, and the force will be directly
> as the versed sine, and inversely as the square of the time.
> Q.E.D.[78]

By virtue of the preceding analyses, and without entering into a
meticulous study of the mathematical rigor of this demonstration,

it follows that if t is the time taken in passing through an arc, then ultimately the versed sine a(t) of this arc is both as the centripetal force in the middle of the arc (by cor. 4, prop. I) and as the square of the time in passing through the arc (by cors. 2 and 3, lem. XI), whence it follows that the centripetal force f in the middle of the arc is as the versed sine and inversely as the square of the time:

$$f \; \alpha \; \frac{a(t)}{t^2}$$

Then, in the first corollary of this proposition, Newton arrives via the law of areas at the general, and geometrically manipulable, expression of central forces:

> Cor. I. If a body P revolving about the centre S describes a curved line APQ, which a right line ZPR touches in any point P; and from any other point Q of the curve, QR is drawn parallel to the distance SP, meeting the tangent in R; and QT is drawn perpendicular to the distance SP; the centripetal force will be inversely as the solid $\frac{SP^2 \cdot QT^2}{QR}$, if the solid be taken of that magnitude which it ultimately acquires when the points P and Q coincide. For QR is equal to the versed sine of double the arc QP, whose middle is P: and double the triangle SQP, or SP · QT is proportional to the time in which that double arc is described; and therefore may be used to represent the time.[79]

In fact, ultimately, QR is equal to the versed sine of the double arc $\overset{\frown}{QP}$, whose middle is P. Additionally, since the doubled area of the triangle SQP is equal to SP times QT, by proposition I (the law of areas) it follows that this area is proportional to the time in which the double arc $\overset{\frown}{QP}$ was described. Thus the time can be rep-

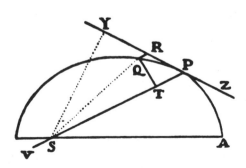

resented geometrically by that area, and through substitution, the centripetal force in P can be expressed, when P and Q coincide, by $f(\text{in P}) \propto \dfrac{QR}{SP^2 \cdot QT^2}$: that is, this force is also "inversely as the solid [i.e., the magnitude which has the dimension of a volume] $\dfrac{SP^2 \cdot QT^2}{QR}$, if the solid be taken of that magnitude which it ultimately acquires when the points P and Q coincide."

Newton now disposed of a geometrically manipulable expression of centripetal force in a point, which enabled him to resolve a number of problems relating to central forces—that is, to obtain at last the expression of the variation of force as a function of the distance between a moving body and a given center of force. Despite the general form of the expression, the solution to each problem could not be obtained "mechanically," as would prove to be the case with the algorithms of the differential calculus, for example, or with Binet's formula. Arriving at a solution required paying specific attention in each case to the particular geometric properties involved, so that by means of the method of first and last ratios a finite equivalent of the general expression could be found. Thus it became possible to determine a relation depending only on the distance SP and constants, when P and Q coincide.

The treatment of the first example given by Newton, in proposition VII, is extremely illuminating:

> If a body revolves in the circumference of a circle, it is proposed to find the law of centripetal force directed to any given point.[80]

Let there be a circle QPA, S being the given center of force, P the place where the body is supposed to be at a given instant, and Q a position next to P on the circle. The tangent to the circle in P is PRZ. How were the QT and QR of the general expression as a function of finite magnitudes to be found, such that this general expression of centripetal force would have meaning when P and Q coincide? The use of the prop-

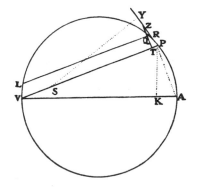

erties of the figure and, particularly, of similar triangles proved to be a great help. To this end, Newton proceeded as follows: "Through the point S draw the chord PV, and the diameter VA of the circle; join AP, and draw QT perpendicular to SP, which produced, may meet the tangent PR in Z; and lastly, through the point Q, draw LR parallel to SP, meeting the circle in L, and the tangent PZ in R."[81] The triangles ZQR, ZTP, and VPA are similar. Consequently,

$$\frac{RP}{QT} = \frac{VA}{VP}.$$

Given that $RP^2 = RQ \cdot RL$ (i.e., the power of a point in relation to a circle), squaring and substitution yields the following:

$$\frac{RQ \cdot RL}{QT^2} = \frac{VA^2}{VP^2} \quad \text{whence} \quad QT^2 = \frac{RQ \cdot RL \cdot VP^2}{VA^2}.$$

It suffices then to multiply both sides of this equation by SP^2/QR to obtain the desired expression of the force:

$$\frac{QT^2 \cdot SP^2}{QR} = \frac{RL \cdot VP^2 \cdot SP^2}{VA^2}.$$

When P and Q coincide, PV and RL are ultimately in the ratio of equality, whence

$$\frac{QT^2 \cdot SP^2}{QR} = \frac{PV^3 \cdot SP^2}{VA^2}.$$

Therefore, "the centripetal force is inversely as

$$\frac{SP^2 \cdot PV^3}{AV^2},$$

that is (because AV^2 is given), inversely as the product of SP^2 and PV^3."[82] The expression of the force is thus indeed given in finite terms, and as a result, the problem is solved. In the particular case (see cor. 1, prop. VII) where the center of force S is placed on the circumference of the circle, at V, the force is "inversely" as PV^5/AV^2, which is to say that it varies as the inverse of the fifth power of the distance PV. The treatment of the problems of central forces was now only a question of geometrical skill—something Newton assuredly did not lack!

Having clearly laid out his principles, Newton succeeded in creating the mathematics of the ultimate, or of the first and last ratios as he called it, which was necessary for working out his theory of central forces; and although the modern reader may be unsatisfied with regard to certain points, it must be granted that he managed to carefully and rigorously demonstrate each of the stages of his construction. Although the requirement of mathematicity was fully affirmed, the investigation of infinity and motion was far from finished. Other techniques were emerging or were being developed in other workshops, and other routes of inquiry were being followed.

THREE

The Science of Motion in the Workshops of Infinity

1. Satisfying Reason

As we indicated in the first chapter, the new era of the science of motion began, strictly speaking, with the work published by Galileo in 1638 in the *Discorsi*.[1] Galileo's fresh approach to the problem of motion bids us to take our leave of Huygens[2] and Newton[3] and return to the opening pages of the third day of the *Discorsi*. It is here that the future of a science which held the promise of mastering the problems of infinity first began to be played out—that is to say, a science which was likely to be able to satisfy reason without being unfaithful to mathematical novelty, and to make possible the deployment of a mathematical physics that would be fully conscious of its inductive power.

In the opening pages of the third day of the *Discorsi*, the concept of uniform rectilinear motion is worked out initially in an axiomatic form inspired by Euclid:

Axiom I

During the same equable motion, the space completed in a longer time is greater than the space completed in a shorter time.

Axiom II

The time in which a greater space is traversed in the same equable motion is longer than the time in which a smaller space is traversed.

Axiom III

The space traversed with greater speed is greater than the space traversed in the same time with lesser speed.

Axiom IV

The speed with which more space is traversed in the same time is greater than the speed with which less space is traversed.[4]

In axioms III and IV, Galileo introduces the velocity of a body in uniform motion as a scalar magnitude. However, this magnitude is not given any explicit definition. In reality, the sole aim of the axiomatic framework was to permit Galileo to apply the Euclidean definition of proportional magnitudes to times and spaces (axioms I and II) and, independently, to velocities and spaces (axioms III and IV), subject to the law of homogeneity (which forbids mention of quantities of different orders of magnitude in the same expression). Credit for moving beyond this theory of proportions was largely due to Descartes, who in the *Regulae ad directionem ingenii* (written in 1628–1629)[5] and then in the *Géométrie* (published in 1637)[6] considered the different kinds of magnitudes—the root, square, cube, and so forth—as nothing more than magnitudes constituting the terms of a single continuous proportion. He went on to assign a unit line or surface to this continuous proportion, which made it possible finally to imagine the different kinds of magnitudes in the same form, whether of a line or of a surface:

> We should note also that those proportions which form a continuous sequence are to be understood in terms of a number of relations; others endeavor to express these proportions in ordinary algebraic terms by means of many dimensions and figures. The first of these they call "the root," the second "the square," the third "the cube," the fourth "the square of the square." I confess that I have for a long time been misled by these expressions. For, after the line and the square, nothing, it seemed, could be represented more clearly in my imagination than the cube and other figures modelled on these. Admittedly, I was able to solve many a problem with the help of these. But through long experience I came to realize that by conceiving things in this way I had never discovered anything which I could not have found much more easily and distinctly without it. I realized that such terminology was a source of conceptual confusion and ought to be abandoned completely. For a given magnitude, even though it is called a cube or the square of a square, should never be represented in the imagination otherwise than as a line or a surface, in accordance with the preceding Rule. So we must note above all that the root, the square, the cube, etc. are nothing but magnitudes in continued proportion which, it is always supposed, are preceded by the arbitrary unit mentioned above.[7]

It was therefore in response to Euclidean requirements as they appeared "prior" to Descartes's *Géométrie* and *Regulae* that Galileo proposed his axiomatic framework for uniform rectilinear motion.

In the case of uniformly (naturally) accelerated motion, velocity, conceived as a continually varying quantity, could no longer support the nonexplicit definition of velocity implied by the statement of axioms III and IV of uniform motion. It must however be acknowledged that Galileo, in the note that follows the definition of uniform motion, envisaged the need to consider "any [times] whatever":

> NOTE: To the old definition, which simply calls motion "equable" when equal spaces are completed [*transiguntur*] in equal times, it seems good to add the qualifier "any whatever," that is, in all times; for it may happen that a moveable passes through equal spaces in some equal times although the spaces completed in smaller parts of those same times, themselves equal, are not equal.[8]

Nonetheless it remained the case, as this passage confirms by its lack of clarity, that a new definition of velocity was needed, capable of being applied to uniformly accelerated (and, more generally, to uniformly varied) motion. For this purpose, Galileo devised the concept of degree of velocity. While to a certain extent it prefigured the concept of instantaneous velocity, it nonetheless remained subject to the Galilean way of conceiving motion, which regarded velocity as an "intensive magnitude" increasing by successive additions of degrees. This concept of degree of velocity, like that of velocity, was not the object of an explicit definition. How did it nonetheless come to have a place in Galileo's deductive analysis of the mathematics of uniformly (naturally) accelerated motion?

In this connection let us consider first of all the opening lines of theorem I, proposition I, of the third day of the *Discorsi*—a theorem and proposition that state the result on the basis of which the law of proportionality between spaces and the squares of the times could be formulated:

Proposition 1. Theorem 1

The time in which a certain space is traversed by a moveable in uniformly accelerated movement from rest is equal to the time in which the same space would be traversed by the same moveable

carried in uniform motion whose speed is one-half the maximum
and final degree of speed of the previous, uniformly accelerated,
motion.[9]

Let us now examine the first lines of the theorem:

> Let line AB represent the time in which the space CD
> is traversed by a moveable in uniformly accelerated
> movement from rest at C. Let EB, drawn in any way
> upon AB, represent the maximum and final degree
> of speed increased in the instants of the time AB. All
> the lines reaching AE from single points of the line
> AB and drawn parallel to BE will represent the in-
> creasing degrees of speed after the instant A.[10]

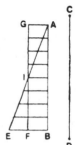

In this passage, Galileo introduces the degree of
velocity by a geometrical representation: a straight-
line segment (represented by "the [equidistant] lines [. . .] parallel
to BE"). How does he then manage to pass from velocity represented in this way to what constitutes the keystone of this theorem,
namely, the comparison of the distances traversed?

> Next, I bisect BE at F, and I draw FG and AG parallel to BA and
> BF; the parallelogram AGFB will [thus] be constructed, equal to
> the triangle AEB, its side GF bisecting AE at I.
>   Now if the parallels in triangle AEB are extended as far as
> IG, we shall have the aggregate of all parallels contained in the
> quadrilateral equal to the aggregate of those included in triangle
> AEB,[11] for those in the triangle IEF are matched by those con-
> tained in triangle GIA, while those which are in the trapezium
> AIFB are common. Since each instant and all instants of time AB
> correspond to each point and all points of line AB, from which
> points the parallels drawn and included within triangle AEB rep-
> resent increasing degrees of the increased speed, while the paral-
> lels contained within the parallelogram represent in the same way
> just as many degrees of speed not increased but equable, it ap-
> pears that there are just as many momenta of speed consumed in
> the accelerated motion according to the increasing parallels of
> triangle AEB, as in the equable motion according to the parallels
> of the parallelogram GB.[12] For the deficit of momenta in the first
> half of the accelerated motion (the momenta represented by the
> parallels in triangle AGI falling short) is made up by the mo-
> menta represented by the parallels of triangle IEF.
>   It is therefore evident that equal spaces will be run through
> in the same time by two moveables, of which one is moved with
> a motion uniformly accelerated from rest, and the other with
> equable motion having a momentum one-half the momentum of

the maximum speed of the accelerated motion; which was [the proposition] intended.[13]

To compare the distances traversed, Galileo therefore makes an argument relying on the comparison of two aggregates of parallels contained in equal figures: the parallelogram AGFB and the triangle AEB.

Moreover, the aggregate of all the parallels contained in the triangle AEB represents the aggregate of all the degrees of speed, or velocity, of a uniformly accelerated motion while the aggregate of all the parallels contained in the parallelogram AGFB represents the aggregate of all the degrees of velocity of a uniform motion. Consequently the agggregates of the degrees of velocity in both motions are the same:

> [. . .] the parallelogram AGFB will [thus] be constructed, equal to the triangle AEB, its side GF bisecting AE at I.
> Now if the parallels in triangle AEB are extended as far as IG, we shall have the aggregate of all parallels contained in the quadrilateral equal to the aggregate of those included in triangle AEB [. . .].[14]

How could one now move, then, from the comparison of these aggregates to the comparison of distances covered? For Galileo, the way leading from one to the other was "evident." In fact, it was evident only on the condition (not explicitly satisfied by Galileo's demonstration) that the mathematical relation in which the aggregates of the degrees of velocity stand to the distances traversed was specified[15] such that if this relation were one of proportionality, then (and only then) it would be possible to conclude from the equality of the aggregates the equality of the distances traversed,[16] and therefore that, indeed, "equal spaces will be run through in the same time by two moveables, of which one is moved with a motion uniformly accelerated from rest, and the other with equable motion having a momentum one-half the momentum of the maximum speed of the accelerated motion."

This result having been established, the second theorem of the third day of the *Discorsi* was very quickly deduced within the framework of Euclidean geometry:

Proposition II. Theorem II

If a moveable descends from rest in uniformly accelerated motion, the spaces run through in any times whatever are to each other as the duplicate ratio of their times; that is, as are the squares of those times.

Let the flow of time from some first instant A be represented by the line AB, in which let there be taken any two times, AD and AE. Let HI be the line in which the uniformly accelerated moveable descends from point H as the first beginning of motion; let space HL be run through in the first time AD, and HM be the space through which it descends in time AE. I say that space MH is to space HL in the duplicate ratio of time EA to time AD. Or let us say that spaces MH and HL have the same ratio as do the squares of EA and AD.

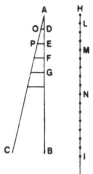

Draw line AC at any angle with AB. From points D and E draw the parallels DO and EP, of which DO will represent the maximum degree of speed acquired at instant D of time AD, and PE the maximum degree of speed acquired at instant E of time AE. Since it was demonstrated above that as to spaces run through, those are equal to one another of which one is traversed by a moveable in uniformly accelerated motion from rest, and the other is traversed in the same time by a moveable carried in equable motion whose speed is one-half the maximum acquired in the accelerated motion, it follows that spaces MH and LH are the same that would be traversed in times EA and DA in equable motions whose speeds are as the halves of PE and OD. Therefore if it is shown that these spaces MH and LH are in the duplicate ratio of the times EA and DA, what is intended will be proved.

Now in Proposition IV of Book I [. . .] it was demonstrated that the spaces run through by moveables carried in equable motion have to one another the ratio compounded from the ratio of speeds and from the ratio of times. Here, indeed, the ratio of speeds is the same as the ratio of times, since the ratio of one-half PE to one-half OD, or of PE to OD, is that of AE to AD. Hence the ratio of spaces run through is the duplicate ratio of the times; which was to be demonstrated.[17]

Strictly speaking, however, the fundamental statement of this second theorem, which was to play such an important role in the development of the science of motion in the seventeenth century, remained unsatisfactory owing to a certain persisting imprecision, in the first theorem, in passing from the comparison of the aggregate of all the degrees of velocity to that of the distances traversed. The recognition of this type of difficulty and the tenacious and unceasing effort to resolve it were solely the result of a real requirement of total mathematicity and of a fully conscious will to construct a mathematical physics.

It is in this sense that Torricelli's attempt to resolve the difficulty, recorded in certain of his manuscripts and revelatory of the general orientation of thought among scientists in the seventeenth century, is to be understood. What one finds in Torricelli's manuscripts is a demonstration resting on the inequality of the points run through in a single instant, depending on whether the motion is uniform or accelerated, the object of which is to compare the relation between the areas of certain figures and the relation between the quantities of all the points of the spaces traversed:

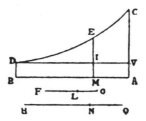

Let BA be the time, and during this time BA let a body traverse the lines GF and OH—on the one hand, GF with a uniform motion having a constant degree of velocity AV, and on the other hand, OH with a nonuniform motion having degrees of velocity homologous to the lines AC or ME.

I say that the spaces traversed GF and OH are between them as the figures BAVD and [BACD].

For there are as many points in the space GF as in the space OH, namely, as many as there are instants in the same time, but these points are unequal.

Now let us take any instant of this time, for example M, and let N and L be the points traversed during this instant [*puncta peracta hoc instanti*]. As the lines MI and ME are to one another, that is to say, as the impetuses [*impeti*], so will be the spaces N and L, and this in all cases.

But the antecedents [. . .] are all equal: consequently, as BAVD is to [BACD], so will be the quantity of all the points of GF to the quantity of all the points of OH, which are the same in number, or in other words so will be GF to OH.[18]

Torricelli supposes therefore that the point N moves along OH in a nonuniform motion with a degree of velocity given by the ordinate ME of the curve DEC in the same time as the point L moves along GF in a uniform motion with a constant degree of velocity given by the constant ordinate AV, or MI of the line DV parallel to BA.

In an instant of time M, the body L travels through the point (or elementary space) L on the line GF and the body N through the point (or elementary space) N on the line OH. These unequal points (or elementary spaces) L and N, each of which must then be supposed to have been crossed with its own degree of velocity, will be to each other, for every instant M, as the degrees of velocity with which each of these points was traversed:

$$\frac{L}{N} = \frac{MI}{ME}$$

Additionally, the motion of L being uniform, all the elementary spaces L and all the degrees of velocity (MI or AV) are equal. This being so, it may be deduced that the quantity of all the points L (i.e., GF) is to the quantity of all the points N (i.e., OH) as all the MI (i.e., BAVD, the Galilean aggregate of all the degrees of velocity MI) are to all the ME (i.e., BACD, the Galilean aggregate of all the degrees of velocity ME). Given this much, it may be said that the relation among the areas BAVD and BACD is as the relation among the spaces traversed and thus, on the basis of this proportionality, that it becomes possible to reason geometrically both about the relation among these areas and about the relation among the spaces traversed.[19]

Torricelli's innovative analysis, whose mathematical daring was similar in many respects to that of Huygens's speculations, examined in the first chapter, was followed up some decades later by Varignon. Varignon's analysis was based on the introduction of a general principle, founded in reason and—by virtue of this—capable of making Galileo's demonstration coherent and rigorous.

Thus Varignon came forward, at the session of the Royal Academy of Sciences on Saturday 19 January 1692, to "demonstrate" the "opinion of Galileo regarding the spaces that falling bodies traverse."[20] "Demonstrate" needs to be construed here in a strong sense, since for Varignon it was a matter of justifying Galileo's result and demonstrating it "in general," that is to say, of giving it a degree of demonstrative perfection identical to that in which Euclidean geometry appeared to be cloaked:

Galileo supposes that the speeds of the bodies that fall increase as the times of their fall; and from that he found that the spaces through which these bodies pass in falling obey the ratio of the squares of the times [the bodies] use up in passing through them, but he proved this only by induction, and not in general. Here is how he could have done it, even following his own principles.[21]

Since Varignon announced his intention of "demonstrating" Galileo's result using the same principles as Galileo, how would he manage to justify Galileo's imprecision?

Varignon's demonstration begins with a geometrical construction which for the most part reproduces that of the *Discorsi*. However, the separate representation of the space traversed is absent from it, as in the demonstration relating to the fall of heavy bodies given by Galileo in 1632 in the *Dialogo*.[22]

It is important, therefore, before going any further, to recall the general framework of the demonstration given in the *Dialogo*. This demonstration, while concealing (as we shall see) an imprecision similar to the one which we have pointed out in the *Discorsi*, nonetheless differs from it somewhat in its structure and vocabulary, so that the demonstration in the *Dialogo* stands in a certain sense nearer to the one given by Varignon in 1692.

In the *Dialogo*, Galileo prefaces his demonstration, as in the *Discorsi*, by saying that the variation of the degrees of speed is continuous, with no small jumps or pauses (see the next section):

> For the increases in the accelerated motion being continuous, one cannot divide the ever-increasing degrees of speed into any determinate number; changing from moment to moment, they are always infinite.[23]

Galileo thus represents each degree of speed by a small line segment such that in the course of the first time AD there is, owing to the asserted continuity of the motion, an infinity of small segments in the triangle AHD:

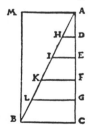

Hence we may better exemplify our meaning by imagining a triangle, which shall be this one, ABC. Taking in the side AC any number of equal parts AD, DE, EF, and FG, and drawing through the points D, E, F, and G straight lines parallel to the base BC, I want you to imagine the sections marked along the side AC to be equal times. Then the parallels drawn through the points D, E, F, and G are to represent the degrees of speed, accelerated and increasing equally in equal times. Now A represents the state of rest from which the moving body,[24] departing, has acquired in the time AD the velocity DH, and in the next period the speed will have increased from the degree DH to the degree EI, and will progressively become greater in the succeeding times, according to the growth of the lines FK, GL, etc. But since the acceleration is made continuously from moment to moment, and not discretely [intercisamente] from one time to another, and the point A is assumed as the instant of minimum speed (that is, the state of rest and the first instant of the subsequent time AD), it is obvious that before the degree of speed DH was acquired in the time AD, infinite others of lesser and lesser degree have been passed through. These were achieved during the infinite instants that there are in the time DA corresponding to the infinite points on the line DA. Therefore to represent the infinite degrees of speed which come before the degree DH, there must be understood to be infinite lines, always shorter and shorter, drawn through the infinity of points of the line DA, parallel to DH.[25]

Galileo represents this infinity of segments (note that he does not specify whether or not they have a thickness) by the surface of the triangle: "This infinity of lines is ultimately [in ultimo] represented here by the surface of the triangle AHD."[26] This surface, quite obviously, does not represent the space passed through, for as Galileo notes: "Thus we may understand that whatever space is traversed by the moving body with a motion which begins from rest and continues uniformly accelerating, it has consumed and made use of infinite degrees of increasing speed corresponding to the infinite lines which, starting from the point A, are understood as drawn parallel to the line HD and to IE, KF, LG, and BC, the motion being continued as long as you please."[27]

What interpretation is to be given to the triangular surface representing all the degrees of speed? Or, more precisely, how is it possible to pass from the totality of the degrees of speed to the spaces traversed?

Here again, as in the *Discorsi*, Galileo's move is to compare a uniform motion and a uniformly accelerated motion. For this purpose he extends all the segments so that they represent the degrees of speed of a uniform rectilinear motion:

> Now let us complete the parallelogram AMBC and extend to its side BM not only the parallels marked in the triangle, but the infinity of those which are assumed to be produced from all the

points on the side AC. And just as BC was the maximum of all
the infinitude in the triangle, representing the highest degree of
speed acquired by the moving body in its accelerated motion,
while the whole surface of the triangle was the sum total of all
the speeds [*la massa e la somma di tutta la velocità*] with which
such a distance was traversed in the time AC [. . .].[28]

And so, far from being the representation of the distance traversed, the surface of the triangle ABC is "the sum total of all the
speeds" corresponding to the passage through a certain space (represented alongside the triangular figure in the *Discorsi;* see the figure above) during the time AC.

The same holds true, then, for the "parallelogram" generated by
the uniform motion:

> [. . .] so the parallelogram becomes the total and aggregate of just
> as many degrees of speed but with each one of them equal to the
> maximum BC.[29]

Having laid out all the elements of his demonstration, it remains
now for Galileo only to compare "the sum total of all the
speeds"—the "aggregates" of the *Discorsi*—of the two motions:

> [. . .] so the parallelogram becomes the total and aggregate of just
> as many degrees of speed but with each one of them equal to the
> maximum BC. This total of speeds is double that of the total of
> the increasing speeds in the triangle, just as the parallelogram is
> double the triangle.[30]

Only at the end of this comparison is the comparison of the
spaces passed through finally introduced; but it is introduced in
terms ("it is indeed reasonable and probable that [. . .]") that are
incapable, to use Fontenelle's expression once more, of fully "satisfying Reason":[31]

> And therefore if the falling body makes use of the accelerated
> degrees of speed conforming to the triangle ABC and has passed
> over a certain space in a certain time, it is indeed reasonable and
> probable that by making use of the uniform velocities coresponding
> to the parallelogram it would pass with uniform motion dur-
> ing the same time through double the space which it passed with
> the accelerated motion.[32]

The assumptions underlying these expressions—underlying the
"it is indeed reasonable and probable" of the *Dialogo* and the "it is
therefore evident" of the *Discorsi*—were what Varignon, in 1692,

wished to make explicit in order to make Galileo's demonstration genuinely rigorous.

We are now ready to revisit Varignon's demonstration, which, in substituting "speeds" for "degrees of speed," begins by presenting a construction that in its main features repeated that of the *Dialogo*:

> Let AB express whatever time one pleases for the fall of a body. Since (by hypothesis) the speeds of this body in falling obey the ratio of the times of its fall, it is evident that if DE expresses the speed acquired in any part AD of this time AB that one pleases, its parallel FG will also express the speed of this body at the end of the time AF because DE is to FG as AD is to AF. For the same reason, HK will express the speed of this body at the end of the time AH;[33] and thus 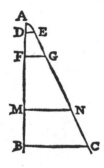 in all imaginable parts of the time AB up to BC, which [taken together] will express the speed of this whole body at the end of this whole time. If therefore through all the points of the line AB one imagines parallels to BC, each of them will express the speed of this body at the end of each of the times expressed by the parts [of] AB, taken from A until each of these points.[34]

Given this, Varignon departs from Galileo in considering, not aggregates, but (in a conceptual shift that risked implicitly reintroducing the problems connected with the composition of the continuum[35] and abandoned Galileo's prior result establishing the equality of certain geometrical figures, specifically of triangle and rectangle) the "sum of all these parallels:"

> Therefore the sum of all these parallels will express the sum of all the speeds that this body has had in all the instants of its fall.[36]

It follows then:

> For example, the sum of all the lines parallel to BC which are in the triangle BAC will express the sum of the speeds that this body has had in all the instants of the time AB; in the same way the sum of these parallels contained in the triangle MAN will express the sum of all the speeds that this body has had in all the instants of the time AM, and thus [too] for the rest.[37]

Varignon, having in a certain sense replaced Galileo's "aggregates" by "sums," can now very simply—but unrigorously—say that "these lines being supposed indefinitely near[38] to each other, it is evident that their sums are as the surfaces of the triangles ABC, AMN, etc."[39] And, as a result, "the sum of the speeds that this body has in falling in the time AB is to [the sum of speeds] it has in falling in the time AM as ABC is to AMN."[40] The two triangles ABC and AMN being similar, the relation of their areas is as that of AB^2 to AM^2,[41] which is to say, as the relation of the squares of the times "used up in falling," whence, finally:

> Thus the sums of the speeds that a body has in all the instants
> of its fall are as the squares of the times that it uses up in falling.[42]

Thus Varignon was able to establish the result that the "sums of the speeds" are as the squares of the times. But how would he be able to show on the basis of this that the spaces are as the squares of the times; that is, how could he construct a relation allowing him to pass from the "sums of the speeds" to the spaces traversed?

In place of the Galilean phrase "it is therefore evident" Varignon introduced a principle, which he regarded as rationally justified ("founded in reason"), namely, that "effects are always proportional to their causes":

> Now [effects being always proportional to their causes], it is
> evident that the spaces that the bodies traverse in falling, are as
> [the] sum of these speeds,[43] [and] so they are also as the squares
> of the times that these bodies use up in falling, [which is] what
> was to be demonstrated.[44]

Varignon borrowed his principle in all probability from John Wallis. The seventh proposition of the first part of Wallis's *Mechanica* (1670–1671) reads: "Effects are proportional to their adequate causes."[45] Wallis goes on to comment on this proposition in a brief scholium: "I have thought that it was necessary to make a premise of this universal proposition since it opens up the path by which one passes from pure mathematical speculation to physics; or rather it links one to the other."[46] Thus this universal proposition made it possible, as Wallis had suggested it might, for Varignon to pass from pure mathematical speculation (by which the sum of all the lines or of all the speeds was obtained) to physics (which is to say to the observation of a space traversed insofar as the space traversed is a physical effect, all the speeds of which constitute the mathematical ratio). For Varignon, the Galilean law therefore was to be interpreted as having passed from the domain of experimental physics into that of mathematical physics, or, as it were, into a domain where the requirement of mathematicity was now fully satisfied.

Three years later, in 1695, an entirely comparable situation—and so a very illuminating one for our purposes—arose in connection with a reappraisal of the status of Torricelli's law.[47] It will be recalled that in 1644, in the *De motu aquarum*,[48] Torricelli had published the law that now bears his name. It stipulates that the speed with which a liquid flows out from the bottom of a tank in which a hole has been pierced is proportional to the square root of the height separating the hole from the free surface of the liquid. This is to say that the speed is equal to that which a drop of water would acquire in free fall from a height equal to the altitude of the water above the drain opening.

The relevant passage in the *De motu aquarum*[49] makes it very clear that Torricelli's work was directly inspired by Galileo's investigations on the fall of heavy bodies. Moreover, the experimental study of this law presented a certain number of difficulties which Torricelli frankly acknowledged, noting that experience seemed both to tell for and against the principle.[50]

This is the reason why some years later, at the Royal Academy's session of 11 July 1668, Picard and Mariotte were charged with responsibility for conducting a study "of the force of running waters in pushing and moving."[51] Two weeks later, on 25 July, Picard announced a specific program of research, insisting on the need "to test what Torricelli said":

> [. . .] the second [aim] is to test what Torricelli said, [namely,] that water which drains out of a vessel pierced on the bottom or the side has in exiting the same speed that it would have acquired in falling from the full height of the water which is in the vessel.[52]

In the course of this same Wednesday session, on 25 July 1668, Christiaan Huygens was "kindly requested" to take part in Picard's study. A set of experiments was subsequently agreed upon, and, at the Wednesday session of 8 August, Huygens presented various reflections bearing particularly upon the validity of Torricelli's law, remarking in this connection:

> With regard to the third experiment he [Huygens] said that to demonstrate it [in the same way] as all the other propositions of

Torricelli's Treatise it was necessary to suppose an effect of nature, which until now not being able to be demonstrated by reason, but only to be proved by experience, must be taken as [a guiding] principle in this matter. It is that water and the other heavy liquid bodies, in leaving by some opening of the vase that contains them, have a speed capable of making them climb back up as high as their surface [level] in the vase.[53]

And consequently:

From this it follows that this speed is equal to that which a drop of liquid would have [on falling] from the height that its surface is raised above the opening. For since a body acquires by its fall exactly as much motion as is necessary to make it climb back up to the height from which it has descended, and that this liquid has exactly [the amount] of it [needed] to produce this effect, it follows that this motion or this speed which it has on leaving the vessel is equal to that which it would acquire in falling from as high [a level] as the surface of the liquid in the vessel.[54]

The validity of Torricelli's law rested therefore, according to Huygens, on the experimental principle according to which liquid bodies, on passing out of a vessel by an opening in it, possess a speed sufficient to climb back up just as far as the free surface of the liquid. Very often, as Torricelli had already emphasized, the experimental evidence was not entirely satisfactory. Thus Huygens devoted himself at length to examining "what are the causes which produce these apparent diversities." The chief causes Huygens distinguished were the role of air resistance, the effect of the water's falling back upon itself, the "adhesion" of the water to the sides of the vessel, and the way in which it passed out of the vessel.[55]

This theoretical reflection about the experimental basis of Torricelli's law, which gave rise to fresh experiments, continued for a year. As Huygens had already written on 8 August 1668, Torricelli's law had not been able to be "demonstrated by reason, but only proved by experience." It remained therefore to provide the desired demonstration, which became the fundamental and specific task of the nascent mathematical physics.

As in the case of the motion of heavy bodies, it was Varignon who was to try to demonstrate "by reason" the law proposed by Torricelli. At the Academy's Wednesday session of 20 April 1695, the presentation of a paper by Varignon was announced in these terms:

M. Varignon has demonstrated independently of experience that the speeds of the streams of water on their exit are always as the roots of their heights.[56]

Varignon presented his paper three days later, on 23 April 1695.[57] Following Huygens, he stated his objectives at the outset: he intended to give a demonstration of Torricelli's law, without limiting himself to any "principle of experience":

> It is a commonly received opinion that the speed of the streams of water at their exit is always as the roots of the heights of the water above the opening which allows it to escape. And there [lies] the first principle of the science of motion and of the measure of running and flowing waters. However, I know [of] no one who has yet demonstrated it. It is true that it is confirmed by an infinity of experiments which have been made by Messieurs Castelli, Borelli, Guillelmini, and above all by Monsieur Mariotte. But although this matter has been one of the most studied, no one (to my knowledge) has yet been able to find the reason for this principle: thus it is that all those who have treated of it have been obliged to assume it merely as a principle of experience, although they undertake to demonstrate an infinity of other truths which are not less confirmed by experience than this principle.[58]

In the interest of establishing a genuine demonstration, Varignon rejected right away the interpretation given by Torricelli, which relied on the model of the accelerated fall of heavy bodies:

> It is not [necessarily] that the reason for [this principle] is well hidden [from us]: but one is led astray by the resemblance that this speed has with that which would result from the accelerated fall of the water, from its surface to the opening by which it departs: for having regarded [the fall] as the effect of such an acceleration, one is naturally led to seek the reason for it in this direction. I have also followed [in this direction] for some time, but having not found anything either, it finally occurred to me that this way, [as] completely natural as it may appear, might well however not be that of nature.[59]

Varignon proposed, therefore, a new approach:

> It is this that led me to examine still more closely what happens in a tube when the water flows out of it; and it seemed to me that the water being contiguous throughout [the] entire length [of the tube], that at the top descended as fast as that at the bottom; and that as a result there was no acceleration in this tube.[60]

In a tube continuously filled with water,[61] the water flowing out from the bottom is subject to "no acceleration" in the sense that its parts are considered as being without relative motion to each other. And, as a consequence:

> It is without doubt because of this that the water flows out not only from this tube, but also from any other vessel, reservoir, or canal as quickly at the beginning as afterward, so long as it remains at the same height in it. This uniformity of speed thus [being] recognized, I have sought the reason of the principle in question in that of uniform motions.[62]

Varignon supposed that, during a small interval of time, one could both neglect the variation in the level of the free surface (in particular, if the section of the vessel is very much higher than that of the hole) and consider the motion as roughly uniform, or (to use a more modern phrase) permanent for the liquid as a whole.

Varignon then provided three demonstrations, distinct from each other but characterized by very similar approaches. The first two made appeal to an earlier paper dated 31 December 1692 concerning rectilinear motions[63] and tried, without great conceptual or technical rigor, to relate the weight of the columns of liquid to the speeds at which they flowed out from a small hole at the bottom and, as a result, to establish the proportionality between the outflow rate and the square root of the height of the liquid above the hole. All that was far from being satisfactory. By contrast, the third solution merits review and presentation here in detail.

In this solution Varignon avoided recourse to previous results and laid out his demonstration in full:

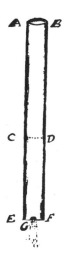

> But without resorting to this general rule, it is [well] enough known that causes are always proportional to their effects; and that in consequence the quantities of motion are always proportional to the motive forces. Now the motive forces here are the weights of the columns of water AF, CF; and the quantities of motion caused by their pressure are as the products of the quantities of water that they push out [through the hole G] in equal times, each multiplied by its speed (that is, in the ratio composed of [the ratios] of their masses and speeds). Therefore, the weights of the columns of water AF, CF (that is, these columns themselves) are in the ratio composed of [the ratios] of the quantities of water that they push out in equal times through the

hole G and of the speeds of these same quantities of water. But since these constituent ratios are equal, because these quantities of water are to each other as the speeds with which they exit the tube AF, the composite [ratio] will be as the square of each one [of them]. Therefore, the columns of water AF, CF, or (what comes to the same [thing]) their heights AE, CE, will be to each other as the squares either of their masses or of their speeds. And consequently either these masses or these speeds, or rather both, will be to each other as the roots of the heights AE, CE of the surface of the water above the opening that lets it escape, which is the principle that needed to be demonstrated.[64]

The principle according to which "causes are always proportional to their effects" is introduced here, as in the treatise on the fall of heavy bodies, to give the demonstration a basis that fully satisfies reason, and so permits an escape from the reign of the empirical and the pragmatic.

Moreover, in Varignon's posthumously published *Nouvelle mécanique ou statique dont le projet fut donné en 1687* (1725), the principle according to which "causes are always proportional to their effects" appears as the first in a list of eight axioms:

> Axiom I. Effects are always proportional to their causes or producing forces, since [the latter] are causes only as far as [the former] are effects, and only by virtue of what they cause in them.[65]

In the same way, in the *Traité du mouvement* (also published posthumously in 1725), Varignon says:

> Axiom II. Effects are proportional to their causes, since [the latter] are causes only as far as they produce effects.[66]

In thus laying down this principle or axiom, Varignon's demonstration in the paper of 1695 rests on the idea that the small quantity of water that escapes the receptacle at each instant receives its entire motion from the pressure exerted equally by the "weights of the columns of water" at the opening.

These weights must be proportional at each instant to the quantity of motion generated in the small quantity of water that flows out. Moreover, the quantities of motion are proportional to the speed and to the mass of the water drained off through the small opening. Now, since the masses of water "are to each other as the speeds with which they exit the tube," as a result the quantities of motion are as the squares of the speeds; and therefore, the heights, which are as the weights (which themselves in turn are as the quantities of motion), are as the squares of the speeds—which suffices to yield Torricelli's law.

This extremely interesting demonstration remains nonetheless very unsure of itself with regard to the handling of infinitesimals in connection with its use of force and quantity of motion. As Jean-Étienne Montucla (1725–1799) was to remark a century later:

> There is nonetheless something vague about this argument, for [it] tacitly supposes that the small mass which escapes from the vessel at each instant abruptly acquires the whole of its speed through the pressure of the column [of water] that meets the opening.
> Now, it is known that [such] a pressure cannot immediately produce a finite speed [. . .].[67]

Montucla went on to develop Varignon's argument, emphasizing the limited character of the demonstration, which was restricted to the case of quasi-stagnant water.

Varignon was to come back to this question eight years later, in a paper delivered to the Academy at its Wednesday session of 14 November 1703.[68] The new paper was intended mainly to generalize the results of the 1695 paper and to isolate a certain number of rules whose particular cases exactly corresponded to the experimental results obtained by his contemporaries. By way of conclusion, Varignon added:

> This general rule & the others of art[icle] 29 will likewise furnish all those of Castelli, Torricelli, Borelli, Guillelmini, etc. [along] with an infinity of others, making use of them as [use] has just been made of it.[69]

In fact, in this paper the generalization resulted from the solving of two problems. The first involved the problem of the outflow of water through a horizontal opening made at the bottom of a vessel, applying, in the context of the demonstrations, the notions of specific gravity (volume weight)[70] and density[71]:

> To find the relations of the speeds of waters & of other liquids of different specific gravity, at [the moment of] their exit [from a vessel] by any horizontal openings whatever, above which these liquids may be at such heights as one likes.[72]

The second problem aimed at a broader generalization since, among other things, the openings may be of any size:

To find a Rule of the motion of waters or of any other liquids whatever that comprehends their specific gravities, the openings or sections by which they flow out, their heights above these openings, the times or durations of their outflows, their discharges or [the quantities of] what flows out during these times, etc., whether one takes these effluxes of liquids as the masses of what flows out during these same times, or one takes them as the volumes of these masses, or finally whether one takes them as the absolute gravities of these same masses." [73]

The arguments advanced by Varignon followed those of 1695 for the most part while introducing certain new elements, borrowed both from the differential calculus to express the discharge as a mass:

> & consequently $\dot{m}$, $\dot{\mu}$, (differentials marked in the manner of M. Newton, the letters d & δ being used subsequently to express the densities of the liquids contained in the tubes ABCD & MOPQ) which flow out with such speeds at the same instant through these openings; [74]

and from the integral calculus to express the surface of the hole through which the water flows out (Varignon actually uses the integration sign in this case, though not as rigorously as one might wish).

As we have noted, Varignon had stressed earlier, in 1695, that Torricelli's law, though "confirmed by an infinity of experiments," had not yet been "demonstrated." Here, in the 1703 paper, he returned to this idea and developed it further:

> It is true that the truth of this proposition is confirmed by an infinity of experiments done by Majottus, Castelli, Torricelli, Borelli, Guillelmini, & above all by M. Mariotte, [who together] approached it from all sides. But for want of having found the reason for it, all those (at least whom I know) who have treated of [this matter] until now, have been obliged to assume it solely as a principle of experience: they have only believed it, I say, on [the strength of] approximate experiments, & which moreover, [being] founded solely on the testimony of the senses, could never be exact enough to be able to be surely established with any rigorous & geometrical precision; besides an infinity of them would be needed to be able to establish it in general. In [the] matter of exact & precise truths, such as that of the preceding relation of the speeds of the liquids to their outflows, experience can at most only make guesses [about] them by dint of approximation; but it could never establish them to the extent of placing them

completely beyond doubt: only reason may achieve [such cer-
tainty]; & this is what is called demonstration.[75]

The purpose of this paragraph, most of which is missing from
the record of the minutes of the sessions of the Royal Academy,[76]
was therefore to clarify for the reader the current status of Torricelli's law by further developing the similar theses advanced in the
paper of 1695.

For Varignon, in 1695 as in 1703, to deduce Torricelli's law from
the axiom that "causes are always proportional to their effects,"
the principles of mechanics, and the general laws of motion was to
properly demonstrate it. He thus implicitly accorded to mechanics
a demonstrative perfection identical to that which Euclidean geometry seemed to enjoy.[77] On this view, to integrate Torricelli's law
with mechanics amounted in fact to finding "the reason for it,"
and so placing it "completely beyond doubt." To quote Fontenelle's
words of 1703 once more, it amounted to "satisfying Reason":

> This relation of speeds will therefore no longer be a [mere]
> principle of experience, and Reason, which no longer has any
> uncertainty to fear, nor anything to beware of, is fully satisfied.[78]

Reason was satisfied by virtue of the fact that the logical requirement of total mathematicity was fully met, or at least met within
a definite mathematical framework, for although the requirement
of mathematicity was regarded as a regulating principle, it too, by
its very nature, was subject to the development of mathematics.
Reason, after all, was satisfied only once it was believed to be satisfied.

2. Ratios of the Beginnings, Ends, and
Continuous Evolution of Motions

Bonaventura Cavalieri (1598–1647), in a letter to Galileo dated 21
March 1626, fully emphasized the importance and the difficulty of
the problems involved in trying to understand the beginning and
continuous evolution of motion:

> [. . .] I have managed to compose a modest [paper] on motion,
> as M. Ciampoli [desired]: when it comes to having to prove that
> the moving body, which from rest has to pass to some degree of
> speed or other, must pass [first] through [all] intermediate [de-

grees], I find no reason that reassures me, though it seems to me that generally it may be so [. . .].[79]

The search for a "reassuring" reason was in this case both a research program and a mental attitude: a desire to understand the beginning and continuous evolution of motion—which is also (and above all) to say, a desire to conceive these things in mathematical terms.

The extreme difficulty of these questions arises from the fact that in attacking them one is immediately confronted with the problem of infinity.

Pascal explicitly addressed the problem toward the end of the 1650s in his brief treatise *De l'esprit géométrique*:

> For no matter how fast a motion may be, it is always possible to conceive one that is faster, and then to accelerate this one even more, and so on infinitely without ever attaining any motion so rapid that it cannot become more so. On the other hand, no matter how slow a motion may be, it can still be retarded, and this retarded motion can again be retarded, and so on infinitely without ever attaining such a degree of slowness that we cannot decrease the speed by an infinite number of gradations without reaching the point of rest.[80]

In 1638, in a very similar vein, Galileo (through his friend and spokesman Salviati) had already said in the *Discorsi:*

> Please hear me out. I believe you would not hesitate to grant me that the acquisition of degrees of speed by the stone falling from the state of rest may occur in the same order as the diminution and loss of those same degrees when, driven by impelling force, the stone is hurled upward to the same height. But if that is so, I do not see how it can be supposed that in the diminution of speed in the ascending stone, consuming the whole speed, the stone can arrive at rest before passing through every degree of slowness.[81]

In this passage, like Pascal but in a more precise way, Galileo emphasized the continuity which on his view characterized the increase or decrease of speed in a naturally accelerated motion. Thus, in such a motion a "rising heavy body does not persist for any finite time in any one degree of speed [. . .]."[82] This again amounts to saying, according to Galileo, that in an accelerated or retarded motion a body that departs from rest or that returns to it passes through an infinity of degrees of speed in a space of time which, no matter how small it may be, contains an infinity of instants.[83] In this sense rest can be considered, not as the opposite of motion, but as a limit or particular case of motion.[84]

Leibniz, some years later, in a letter to Varignon dated 2 February 1702, adopted the same approach in speaking of his "law of continuity," by virtue of which

> [. . .] we may consider rest as infinitely small motion (that is, as equivalent to a particular instance of its own contradictory), coincidence as infinitely small distance, equality as the limit of inequalities, etc.[85]

Already in 1687, in a letter to Arnauld, we find Leibniz asserting:

> And it is a defect of the arguments of M. des Cartes and of [Malebranche] not to have considered that everything which is said about motion, inequality, and elasticity must also be verified when one assumes these things [to be] infinitely small, or infinite. In which case (infinitely small) motion becomes rest, (infinitely small) inequality becomes equality, and (infinitely sudden) elasticity is nothing other than extreme hardness; somewhat like the way that all of what the geometers demonstrate about an ellipse is verified [to hold for] a parabola, when [this] is conceived as an ellipse whose other focus is infinitely remote.[86]

Similarly again in 1704, in the preface to the *Nouveaux essais sur l'entendement humain* (not published until 1765 by Raspe), Leibniz remarked:

> Nothing takes place suddenly, and it is one of my great and best confirmed maxims that nature *never makes leaps*. I called this the Law of Continuity when I discussed it formerly in the *Nouvelles de la république des lettres* [. . .]. There is much work for this law to do in natural science. It implies that any change from small to large, or vice versa, passes through something which is, in respect of degrees as well of parts, in between; and that no motion ever springs immediately from a state of rest, or passes into one except through a lesser motion; just as one could never traverse a certain line or distance without first traversing a shorter one.[87]

It was in fact almost twenty years earlier, in an article entitled "Extrait d'une lettre de M. L. sur un Principe Général, utile à l'explication des lois de la nature, par la considération de la Sagesse Divine; pour servir de réplique à la réponse du R. P. M[alebranche]," published in the July 1687 issue of the *Nouvelles de la république des lettres,* that Leibniz introduced his law of continuity, without giving it a name:

> [. . .] a *principle of general order* [. . .] which is of great value in reasoning [. . .]. This principle has its origin in the *infinite* and is absolutely necessary in geometry, but it is effective in physics as well [. . .]. [T]he principle serves me as a test or criterion by which to reveal the error of an ill-conceived opinion at once and from the outside, even before a penetrating internal examination is begun. It can be formulated as follows. *When the difference between two instances in a given series or that which is presupposed can be diminished until it becomes smaller than any given quantity whatever, the corresponding difference in what is sought or in their results must of necessity also be diminished or become less than any given quantity whatever.* Or to put it more commonly, *when two instances or data approach each other continuously, so that one at last passes over into the other, it is necessary for their consequences or results (or the unknown) to do so also.* This depends on a more general principle: that, *as the data are ordered, so the unknowns are ordered also.* [*Data ordinatis etiam quaesita sunt ordinata.*][88]

In fact, although this passage, like that from the letter to Arnauld cited previously, dates from 1687, Leibniz was in possession of his law of continuity some ten years earlier, as a reading of both the manuscript of *De Corporum Concursu* (written in January–February 1678)[89] and his June 1679 letter to Theodor Craanen[90] will show.

However, this assertion of the continuous evolution of motion and of the passage from rest to motion without leaps, particularly stressed by the various statements of the Leibnizian law of continuity, was far from seeming obvious in the seventeenth century. The reminder of Zeno's paradoxes, and the (explicit or implicit) empirical references made to them, were quite often perceived as so many hindrances to the requirement of mathematicity.

Thus Descartes, after having read Galileo's *Discorsi* (particularly the passage just cited), rejected his continualist theses for the most part in his letter to Mersenne of 11 October 1638:

> [When] Galileo says that bodies which descend pass through all the degrees of speed, I do not believe that ordinarily it happens thus, but indeed that it is not impossible that it sometimes happens [thus].[91]

Moreover, Descartes, in stating in the *Principles of Philosophy* his laws of impact,[92] the revision of which depended in Leibniz's opinion on, among other things, the application of his law of continuity,[93] placed himself in direct contradiction to the requirements of continuity.

The first law of impact states that if "two bodies [. . .] were completely equal in size and were moving at equal speeds [toward each other in a straight line], when they collided, they would spring back and subsequently continue to move [. . .] without having lost any of their speed";[94] but if one now follows Descartes, in the second and third laws, in supposing that one of the bodies is "slightly larger"[95] or that it moves "slightly more rapidly,"[96] then only the smaller one (by the second law) or the slower one (by the third law) will bounce back, such that in each case the other body will follow, both toward the same side. Such an analysis was utterly in contradiction to the very idea of continuity, as Leibniz stressed, for example, in his *Animadversiones in partem generalem Principiorum Cartesianorum* (1692):

> [. . .] two instances which have an infinitely small variation in the hypotheses or given conditions (that is, a difference smaller than any given amount) will nevertheless have the greatest and most noticeable difference in their results [. . .]. [. . .] the rule for equal bodies or for bodies with infinitely small inequality could not be subsumed under the general rule for inequality.[97]

In Leibniz's eyes, in view of the fact that his "law of continuity"[98] served as a "general criterion"[99] or "touchstone,"[100] this amounted at the least to manifest incoherence. In fact, for Leibniz, it will be recalled, by virtue of this law "it is impossible for the result of a disappearing inequality not to vanish into the result of equality."[101]

The attachment to discontinualist theses in the seventeenth century can be further illustrated by excerpts from several particularly representative writings. The first is taken from the analysis of varied motion proposed by François Bernier, who reviewed Zeno's arguments regarding Achilles and the tortoise in his *Abrégé de la philosophie de Gassendi* (1678) and concluded that the differences in the speeds of the two moving bodies resulted from the "mixture of a greater or smaller quantity of small rests":

However, as the comparison that is made between the slow- and the fast-moving body may pose some difficulty not only for those who admit mathematical points but also for those who, like us, recognize only physical points, it is fitting to say something about it here. The difficulty consists in the fact that, if the motion of the slow-moving body and that of the fast one [are] continuous, it is nonetheless necessary that the less rapid one move [for] an instant and that it cover an indivisible physical [point], [whereas] the faster one traverses at once and with no succession several indivisible physical [points] ranked in order, which is incomprehensible. To reply to this difficulty, might it not be said that slowness has its origin in rest? Certainly, [just] as we conceive that the light of the sun at midday is very great, and that the different degrees of what we perceive from this moment until the pure darkness [of night] are born of the mixture of a greater or smaller quantity of darkness, we can similarly conceive that the motion by which atoms are carried through the void is very rapid, and that all the other degrees which are [present] from this motion until complete rest are born of the mixture of a greater or smaller quantity of small rests. By this means one goes from the whiteness of milk and snow to the blackness of coal and the crow, from the heat of fire to the cold of ice, and thus of the very rest affirmed by Aristotle, who did not deny that these changes occur by the mixture of their contraries. This is why, when there are two moving bodies, one of which moves twice as fast as the other, it is necessary to conceive that, of two moments, in both of which the faster one moves, the less rapid one moves only in one and rests in the other; and that for a similar reason, when the motion is three times more rapid, in the three moments in which the faster one moves, the less rapid one moves only in one and rests in the two others, and so on. And say not that this sort of movement will therefore not be continuous, for although it will not truly be thus in itself, it will nonetheless be so to the senses; in the same way that the fire of a match, which is lit at the end and rapidly spun around, appears to view to be circular, and the match to be continuously in any part of the circle whatever, [when in fact] it is in [each part] only successively and with interruption.[102]

A second extract is to be found in Mariotte's *Traitté de la percussion ou chocq des corps* (1673).[103] In this passage, Mariotte rejects the idea that an accelerated motion may be so from the first instant. His argument, like so many others of the period, rests on recalling the difficulties generated by Zeno's paradoxes about infinity, but also, despite the fact that the infinitesimal context seemed unfavorable to such experiments, on various experiments relating chiefly to the outflow and force of fluids:

> Galileo makes some rather plausible arguments [in an attempt] to prove that at the first moment that a weight begins to fall, its speed is smaller than any that may be determined: but these arguments are founded on divisions to infinity, as much of speeds as of the spaces passed through, and of the times of fall, [on] which [account these] are very suspect arguments, like that which the ancients made to prove that Achilles could never overtake a tortoise, to which argument it is difficult to reply and give the solution of it; but the falsity of it is demonstrated by experience, and by other arguments easier to conceive. Thus one will object to Galileo [using] the above arguments which are easy to conceive, particularly that of the scale, and which are much clearer than his, which he has founded on divisions to infinity, which are inconceivable, and on certain rules of the acceleration of the speed of bodies, which are doubtful: for one cannot know whether the falling body does not pass through a small space, without accelerating its first motion, because it takes time to produce the majority of natural effects, as it seems when one makes [a piece of] paper pass through a great flame, with a great speed, without its catching on fire; and as a consequence one must prefer the above arguments to those of Galileo.[104]

Later, at the Saturday session of the Royal Academy of Sciences of 31 July 1677, Mariotte noted in the first lemma of a paper on the fall of light bodies:

> Bodies that fall by their own weight in free air begin their fall with a determinate [and] rather considerable speed, and [do] not pass through all the degrees of slowness.[105]

Simple observation had thus been elevated to the status of a lemma, an argument against mathematical reason and a hindrance to the requirement of mathematicity.

One finds an attitude very similar to Mariotte's in a letter addressed by Hartsoeker to Leibniz on 6 January 1712:

> There is a law in Nature, you say, Sir, which implies that there is no passage *per saltum*. I grant you it in a certain sense; but when you say that this law does not permit there to be any medium between the hard and the fluid, I see no necessity in it. If you did not know by experiment, Sir, that a body which is impelled with any amount of speed you like can return to rest from the instant of the shock without losing little by little and by de-

grees its [imparted] motion, would you not say by your law that
this is impossible?[106]

Leibniz's reply, dated 8 February 1712, was scathing:

> To refute the law of nature, which perhaps I published first
> [to the effect] that no passage occurs by [a] leap, you give this
> experiment, that a body can pass in an instant from motion to
> rest in [the aftermath of] the shock, however rapid this motion
> may be. But when you will have a bit of leisure, Sir, to consult
> what I have said about this law in the *Nouvelles littéraires* of M.
> Bayle, you will find that it is this very pretended experiment, and
> other similar [ones], which I have refuted; and when you will
> look closely at the same experiment, you will find that the bodies,
> however hard they may be, submit to and obey the shock, and
> [such a body], losing the force of [its] motion little by little (as an
> inflated balloon would do), which is transferred into the invisible
> parts [that are] productive of elasticity, acquires that same force
> again when it throws back [the body that initially hit it] by means
> of the spring which returns [it] to [its] original state. It is true
> that M. des Cartes, very great man though he was, was himself
> in error [on this point]: but he was generally [in error] on the
> laws of motion."[107]

The treatment of the evolution of motion "without leaps" or
"pauses," to use Mersenne's expression,[108] therefore appeared
throughout the entire seventeenth century and the early part of the
eighteenth century, as the exchanges that we have just looked at
make clear, to be the result of a daring theoretical choice, sometimes opposed to observation and in any case involving Zeno's
paradoxes, but one that was decisive since, as Galileo and Leibniz
had rightly perceived, it was the very possibility of a mathematical
response to the problem of explaining the evolution of motion that
was at stake.

How were the difficulties associated with the treatment of the
evolution of motion "without leaps" or "pauses" to be escaped?
Or rather: could the continuous evolution of motion be mathematized?

2.1. *The Hobbesian Analysis*

Thomas Hobbes (1588–1679), whose scientific work has sometimes been unduly neglected, was the first to sketch the conceptual
basis for the future treatment of the continuous evolution of mo-

tion which Leibniz, around 1670, was to elaborate in a theoretically fruitful way that would prove to be influential.

In the *De Corpore*,[109] in the eighth chapter of the second part of the book, Hobbes defined motion as continuous:

> 10. Motion is a continual relinquishing of one place, and acquiring of another; and that place which is relinquished is commonly called the *terminus a quo*, as that which is acquired is called the *terminus ad quem;* I say a continual relinquishing, because no body, how little soever, can totally and at once go out of its former place into another, so, but that some part of it will be in a part of a place which is common to both, namely, to the relinquished and the acquired places. For example, let any body be in the place ACBD; the same body cannot come into the place BDEF, but it must first be in GHIK, whose part GHBD is common to both the places ACBD and GHIK, and whose part BDIK, is common to both the places GHIK, and BDEF.[110]

```
A G B I E
| : | : |
| : | : |
C H D K F
```

In the following paragraph Hobbes affirms the continuity and existence of motion as against one of the arguments attributed to Zeno by Diogenes Laertius. Diogenes had written:

> Zeno undermines motion by the following argument: that which moves moves neither in a place in which it is, nor in a place in which it is not.[111]

A more developed version of this same argument was also made, for example, by Epiphanius:

> Zeno argues as follows: What moves moves either in a place in which it is, or in a place in which it is not. But it moves neither in a place in which it is, nor in a place in which it is not; therefore nothing moves.[112]

To this argument Hobbes replied by emphasizing, as Galileo had done in the *Discorsi,* that if one accepts that to the continuity of motion there correspond continuities of time and space, it is false to suppose that a body may move in the place in which it is, for it is in a place only when it is at rest, and when it is in motion, it is not in this place, even for some arbitrarily small period of time:

> 11. That is said to be at rest, which, during any time, is in one place; and that to be moved, or to have been moved, which,

whether it be now at rest or moved, was formerly in another place than that which it is now in.

From which definition it may be inferred, first, that *whatsoever is moved, has been moved;* for if it be still in the same place in which it was formerly, it is at rest, that is, it is not moved, by the definition of *rest;* but if it be in another place, it has been moved, by the definition of *moved.* Secondly, that *what is moved, will yet be moved;* for that which is moved, leaveth the place where it is, and therefore will be in another place, and consequently will be moved still. Thirdly, that *whatsoever is moved, is not in one place during any time, how little soever that time be;* for by the definition of rest, that which is in one place during any time, is at rest.

There is a certain sophism against motion, which seems to spring from the not understanding of this last proposition. For they say, that, *if any body be moved, it is moved either in the place where it is, or in the place where it is not; both which are false; and therefore nothing is moved.* But the falsity lies in the major proposition; for that which is moved, is neither moved in the place where it is, nor in the place where it is not; but from the place where it is, to the place where it is not. Indeed it cannot be denied but that whatsoever is moved, is moved somewhere, that is, within some space; but then the place of that body is not the whole space, but a part of it, as is said above in the seventh article. From what is above demonstrated, namely, that whatsoever is moved, has also been moved, and will be moved, this may also be collected, that there can be no conception of motion, without conceiving past and future time.[113]

These ideas are taken up again and summarized several pages further on, in the first article of chapter 15 of part 3 of *De Corpore*, after which point Hobbes will be ready to introduce the most original (and, for Leibniz's future development of the calculus, the most important) of his concepts, namely, that of *conatus:*

> I have already delivered some of the principles of this doctrine in the eighth and ninth chapters; which I shall briefly put together here, that the reader in going on may have their light nearer at hand.
>
> First, therefore, in chap. VIII. art. 10, *motion* is defined to be *the continual privation of one place, and acquisition of another.*
>
> Secondly, it is there shown, that *whatsoever is moved is moved in time.*
>
> Thirdly, in the same chapter, art. 11, I have defined *rest to be when a body remains for some time in one place.*

Fourthly, it is there shown, that *whatsoever is moved is not in any determined place;* as also that the same *has been moved, is still moved, and will yet be moved;* so that in every part of that space, in which motion is made, we may consider three times, namely, the *past,* the *present,* and the *future time.*[114]

The concept of *conatus* ("endeavor") is then introduced in article 2:

[. . .] I define *endeavor to be motion made in less space and time than can be given;* that is, *less than can be determined or assigned by exposition or number;* that is, *motion made through the length of a point, and in an instant or point of time.*[115]

But what exactly is a point for Hobbes? He expresses his view of the matter in the lines that immediately follow:

For the explaining of which definition it must be remembered, that by a point is not to be understood that which has no quantity, or which cannot by any means be divided; for there is no such thing in nature; but that, whose quantity is not at all considered, that is, whereof neither quantity nor any part is computed in demonstration; so that a point is not to be taken for an indivisible, but for an undivided thing [. . .].[116]

On this view, a point preserves an extension and cannot be reduced, as Leibniz was to suppose some years later, to something unextended, which for Hobbes does not exist. *Conatus* in the Hobbesian sense, then, is an undivided (though not therefore indivisible) spatiotemporal point. An instant enjoys a thoroughly similar status:

[. . .] as also an instant is to be taken for an undivided, and not for an indivisible time.[117]

It is important here to return briefly to the Hobbesian concept of a point if we wish to understand the properties of *conatus* and, in particular, if we are to be able to completely grasp its role in the beginning of the continuous evolution of motions and in their end.

In *The Six Lessons to the Professors of Mathematicks,* published in 1656 together with the English translation of *De Corpore,* Hobbes considered the Euclidean definition of a point ("A point is that which has no part")[118] to be extremely ambiguous. Hobbes remarked that to assert this of a point was to say that a point is either indivisible or it is undivided. But, for Hobbes, to say that a point is indivisible was to say that it is nothing:

> That which is indivisible is no quantity; and if a point be not
> quantity, seeing it is neither substance nor quality, it is nothing.
> And if Euclid had meant it so in his definition, as you pretend he
> did, he might have defined it more briefly, but ridiculously, thus,
> *a point is nothing.*[119]

On the other hand, if a point is something not divided, something undivided, it is nonetheless susceptible of division, of being
a quantity; but in this case, then, the Euclidean definition does
not succeed in stating what a point is insofar as it is a point. For
Hobbes, in fact, a point is not that which is without quantity but
that which is considered as being without quantity. One reads in
his *De Principiis et Ratio-cinatione geometrarum* (1666):

> [. . .] a point is something divisible, but not something whose
> parts must be taken into account in a demonstration.[120]

Given this, Hobbes was perfectly free to hold that points have
extension. As a result, they can admit of extension in different
ways, which is why there can also be instances of *conatus* of different sizes—it being understood that one ought not in any fashion,
for purposes of demonstration, take into account the parts of these
different instances of *conatus,* just as one need not take into account the parts of points. Thus in *De Corpore* the various specific
beginnings and endings of motions can finally be linked up with
the different types of *conatus:*

> In like manner, endeavor is to be conceived as motion; but so
> as that neither the quantity of the time in which, nor of the line
> in which it is made, may in demonstration be at all brought into
> comparison with the quantity of that time, or of that line of
> which it is a part. And yet, as a point may be compared with a
> point, so one endeavor may be compared with another endeavor,
> and one may be found to be greater or less than another. For if
> the vertical points of two angles be compared, they will be equal
> or unequal in the same proportion which the angles themselves
> have to one another. Or if a strait line cut many circumferences
> of concentric circles, the inequality of the points of intersection
> will be in the same proportion which the perimeters have to one
> another. And in the same manner, if two motions begin and end
> together, their endeavors will be equal or unequal, according to
> the proportion of their velocities; as we see a bullet of lead de-
> scend with greater endeavor than a ball of wool.[121]

This element of motion, or *conatus,* can also undergo alterations
and transformations, but in all cases its role is not to furnish a

description of observed phenomena but to supply "the reason of things":

> 7. All endeavor, whether strong or weak, is propagated to infinite distance; for it is motion. If therefore the first endeavor of a body be made in space which is empty, it will always proceed with the same velocity; for it cannot be supposed that it can receive any resistance at all from empty space; and therefore (by art. 7, chap. IX) it will always proceed in the same way and with the same swiftness. And if its endeavor be in space which is filled, yet, seeing endeavor is motion, that which stands next in its way shall be removed, and endeavor further, and again remove that which stands next, and so infinitely. Wherefore the propagation of endeavor, from one part of full space to another, proceeds infinitely. Besides, it reaches in any instant to any distance, how great soever. For in the same instant in which the first part of the full *medium* removes that which is next [to] it, the second also removes that part which is next to it; and therefore all endeavor, whether it be in empty or in full space, proceeds not only to any distance, how great soever, but also in any time, how little soever, that is, in an instant. Nor makes it any matter, that endeavor, by proceeding, grows weaker and weaker, til at last it can no longer be perceived by sense; for motion may be insensible; and I do not here examine things by sense and experience, but by reason.[122]

This concept of *conatus,* which, by going beyond experience, or rather sensation, aims at grasping "the reason of things," the reason for motion, for its beginning and for its end, was adopted in large measure by Leibniz some years later in the theory of motion that he worked out in 1670–1671.

2.2. The Leibnizian Approach

Leibniz developed his theory of motion in the *Hypothesis physica nova,* written in 1670 and published the following year.[123] Under this title he brought together two complementary works: the *Theoria motus concreti* ("Theory of Concrete Motion, or Hypothesis about the Ratios of the Phenomena of Our World," to translate its full title)[124] and the *Theoria motus abstracti* ("Theory of Abstract Motion, or Universal Ratios of Motions, Independent of the Sensible World and of Phenomena").[125]

The main purpose of the *Theoria motus abstracti,* as its title suggests, was to build an a priori or purely rational theory of motion, while the *Theoria motus concreti* was intended to treat motion "as it is actually met with in nature." [126] The former work rested on a set of twenty-four "Fundamental Principles" *(Fundamenta praedemonstrabilia)* broadly inspired—directly inspired, in the case of the initial principles—by the work of Bonaventura Cavalieri as presented in his celebrated *Geometria indivisibilibus* (1635).

It needs to be recalled that Cavalieri, contrary to the position that was to be attributed to him by various authors at the end of the seventeenth century (including, as we shall see, Leibniz) when they spoke of the "method of indivisibles," did not take sides in the debate over the composition of the continuum. Cavalieri did not claim that his indivisibles were the constitutive elements of geometrical objects, of infinitesimals. What he did propose, out of a desire to remain within the Euclidean tradition, and so to avoid the problems associated with passing to a limit that infinitestimal methods involved, was essentially a comparison of figures. This comparison was made possible by indivisibles in the sense that, following Cavalieri, the study of the relation between figures could be replaced by the study of the relation obtaining among indivisibles. As for indivisibles themselves, obviously they were not to be arrived at by shrinking the plane to the point that it became infinitely small, or by shortening figures to the point that they became infinitely small lines, but by cutting figures through the parallel intersection of a plane (in the case of volumes) or of a straight line (in the case of planes). In his demonstrations, then, Cavalieri considered infinite aggregates, rather than sums, of all the planes or of all the lines corresponding to the successive positions of the moving plane or line that intersected the figures.

Accordingly, for Leibniz, in his *Theoria motus abstracti*, motion is a continuum: it is "not anywhere interrupted by little intervals of rests," [127] as we noticed earlier in Bernier's commentary on Gassendi. [128] And so, insofar as it is a continuum, motion (which moreover, according to Leibniz, is the characteristic feature of every continuum) not only is divisible to infinity but is in fact divided, in the sense that "[t]here are actually parts in a continuum" and that "these are actually infinite." To these two principles Leibniz adds a third (noting that "the indefinite of Descartes is not in the thing but in the thinker"), namely, "There is no minimum in space or in a body," [129] for such a minimum (in view of the fact that neither size nor part can be nothing, since "such a thing cannot have any position") "implies a contradiction." In this case, there would in fact be as many minima in the whole as in a part of it,[130] since every part is itself, like the whole, infinitely divisible.

Leibniz escapes this contradiction by relying on his interpretation of Cavalieri's method, a method which he is led to consider (no doubt because he was more preoccupied at this point by the analysis of motion and of trajectories than by purely geometric questions) in terms of the problem of the composition of the continuum. Leibniz therefore introduces with his fourth principle the concept of indivisibility: "There are indivisibles or unextended beings, for otherwise we could conceive neither the beginning nor the end of motion or body."[131] A number of passages we have looked at talk about the "beginning" *(initium)*. What in fact was to be said about it (whether it be the beginning of a body, a space, a duration, or a motion)?

> The proof of this is as follows. There is a beginning and an end to any given space, body, motion, and time. Let that whose beginning is sought be represented by the line *ab,* whose middle point is *c,* and let the middle point of *ac* be *d,* that of *ad* be *e,* and so on. Let the beginning be sought at the left end, at *a.* I say that *ac* is not the beginning, because *cd* can be taken from it without destroying the beginning; nor is it *ad,* because *ed* can be taken away, and so forth. So nothing is a beginning from which something on the right can be removed. But that from which nothing extended can be removed is unextended.[132]

Thus, to quote Pierre Costabel: "Such a beginning belongs to space, time and motion without being itself divisible, for the idea of a divisible beginning is contradictory. Consequently there are many indivisibles constituting space, time and motion; nevertheless they are heterogeneous with respect to that which they constitute seeing that extension cannot be 'assigned' to them without our falling into one contradiction from another.[133]

Leibniz did not hesitate, then, to introduce and to argue from the existence of these surprising mathematical entities, which represent "the beginning of body, space, motion, or time—namely a point, conatus, or instant."[134] It was from Hobbes, with whose thought he had just recently become acquainted, that he had borrowed the concept of effort (i.e., *conatus,* or "endeavor"), transforming it (in the form of his tenth principle) into his own concept of an indivisible of motion: "Conatus is to motion as a point to space, or as one to infinity, for it is the beginning and end of motion."[135]

Leibniz goes on to describe his version of *conatus* in somewhat greater detail. In particular, he says, it is necessary not to confuse the beginning or end of a motion with rest (as a consequence of the sixth principle: "The ratio of rest to motion is not that of a point to space but that of nothing to one").[136]

The Leibnizian principle of continuity, associated with the procedures of the new differential calculus, was to have the effect (as we saw in the first citations in this section) of filling up the gap between rest and motion.

It is nonetheless the case that in the *Theoria* of 1670–1671, written prior to the general formulation of his law of continuity, what Leibniz was proposing was a mathematical treatment, so far as one was possible, of an evolution of motion without leaps or small rests. Just the same, the very subtle argument about divisibility and indivisibility comes as a surprise. Thus Simon Foucher, analyzing this early work in the light of the law of continuity,[137] wrote in his letter to Leibniz of March 1693:

> As for that which concerns your Axiom, *Natura non agit saltatim,* I confess to you, Sir, that I would hardly have conceived your sentiment thereof had two treatises not fallen into my hands, the one *De motu concreto,* and the other *De motu abstracto* [. . .]; I confess that I do not understand how you admit divisibles and indivisibles together: for that redoubles the difficulty and does not resolve the question. In fact, to adjust the parts of the time with those of the space that the moving bodies traverse, it is necessary that indivisibility and divisibility be joined together the one with the other. For if an instant, for example, being supposed indivisible, corresponds nonetheless to a divisible point, the first part of this point will be traversed when the instant will still be only half passed; and this being [so], it will indeed be necessary that this instant be able to be divided up, since it will [only] be half used up, before its other part can actually be. The same thing may be said about a point [that is] indivisible in relation to an instant which can be divided. But on the other hand, if one supposes that the instants and points may also be indivisible, one will not be able to resolve the difficulty of the Skeptics nor show how Achilles is to go faster than a tortoise.[138]

To these difficulties came to be added another, connected with the explanation of the diversity of motions. In 1670–1671, Leibniz explained this diversity by that of the types of *conatus*. These could in fact be compared with each other on the condition that one joined Leibniz in assuming that time passes uniformly, that is, that "every instant is equal to every other one" and that indivisibles of time are all taken to be equal.[139] It is therefore the inequality of points (or indivisibles of space) in an instant that accounted for the inequality of efforts and, as a consequence, for the diversity of motions:

> No one can easily deny the inequality of conatuses, but from this the inequality of points follows. One conatus is obviously greater than another, or one body, moving more rapidly than another, obviously passes through more space from the beginning, for if it passes through the same amount at the beginning, it will always continue to pass through the same amount, for motion continues as it begins [. . .].[140]

To which Foucher, pursuing the argument quoted earlier in the same letter of March 1693, concludes:

> Instants and points are absolutely and mathematically divisible, it will be said, but they are not actually divided into all their possible parts; and this [being] posited, a large point and a small [one] are traversed in a single instant. I allow it; but if this is so, nature will act by [a] leap: for there will occur an instantaneous transport from one extremity of a point to the other, for it is supposed that this transport takes place in an instant, and the same difficulty still remains to be resolved.[141]

By 1693, however, Leibniz had managed to overcome the mathematical difficulties implied by the conceptualization of the evolution of motion set out in this early work. He was able therefore to give a very clear reply to Foucher, making reference to the works of Grégoire de Saint-Vincent:[142]

> [. . .] there are several [subjects] on which I believe [myself] to be better instructed at present; and among others I explain myself altogether differently today on [the subject of] indivisibles. That essay [*Theoria motus abstracti*] was the [work] of a young man who had not yet contributed to mathematics. The laws of abstract motion which I had given then would in fact occur [in nature], if in bod[ies] there was nothing other than what is imagined by Descartes, and even by Gassendi. But as I found that nature makes use of [them] altogether differently with regard to motion, this is one of my arguments against the received notion of the nature of bodies [. . .]; as for indivisibles, when one understands

by that [term] simply the extremities of time or of line, one cannot conceive new extremities or parts in them, either actual or potential. Thus, points are neither large nor small, and it is not necessary to [make a] leap to go beyond them. However, the continuum, though it has such indivisibles everywhere, is not composed of them, as the objections of the Skeptics seem to assume, which [objections] in my opinion are not at all insurmountable, as one will find in properly expressing them. Père Grégoire de St Vincent has indeed shown, by the very calculus of the divisibility to infinity, the place where Achilles must overtake the tortoise that [started ahead of] him, according to the proportion of speeds. Thus geometry serves to dispell these apparent difficulties [. . .].[143]

These difficulties were in large part, as Leibniz indicates, "dispelled" by a novel conceptual approach that was essentially mathematical in nature, based upon a reworking of the calculus of infinity, which we will analyze in the following chapter.

The beginning, end, and continuous evolution of motion therefore progressively acquired their full mathematical status in thus satisfying the requirement of total mathematicity that governed the historical development of the science of motion. Important conceptual work remained to be done, however, for what was at stake, as we will see in the Epilogue, was not only the uses to which the Leibnizian calculus could be put but, more profoundly, the clarification of the foundations of this calculus and the conception of science that such an attempt implied.

The problems of the science of motion were now, strictly speaking, mathematical problems. Mathematical physics was unmistakably asserting its presence.

FOUR

Motion Algorithmized

1. Introduction and Import of the Leibnizian Calculus

Although he possessed the main elements of his new calculus from 1676,[1] it was only in the October 1684 issue of the new journal *Acta Eruditorum*[2] that Leibniz published the first statement of it, "Nova methodus pro maximis et minimis" (or, to give it its full title in English, "A New Method for Maxima and Minima, as Well as Tangents, Which Is Not Obstructed by Fractional or Irrational Quantities").[3] This article, with its aim of providing a new method for determining not only maxima and minima but also tangents, free of the difficulties that encumbered existing techniques, introduced in effect a new method for studying curves. Though it was only six pages long, the article nonetheless made very difficult reading for Leibniz's contemporaries not only because of its mathematical novelty but also owing to poor typography and the extreme concision of the author's style. Although Leibniz managed in so short a space to state the chief rules of his calculus, he did so without demonstrating them.

At the outset of the article, in a very brief introductory paragraph, Leibniz defines on the basis of the geometric elements of a figure what he called *differentia,* or "difference." The reader is asked to consider "the axis AX and different curves VV, WW, YY, ZZ," whose "ordinates perpendicular to the axis" VX, WX, YX, and ZX are "called, respectively, v, w, y, z."

The segment AX marked off along the axis is called x. Next Leibniz introduces the tangents VB, WC, YD, and ZE to the different curves, tangents which intersect the axis AX at B, C, D, and E, respectively, and, finally, defines his concept of "difference":

Let then an arbitrarily cho-sen straight line segment be called dx, and [let] dv (dw, dy, or dz), that is to say, the difference [*differentia*] of v (of w, of y, or of z), be a segment which is to dx as v (w, y, or z) is to XB (XC, XD, or XE).[4]

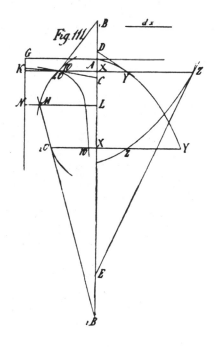

Given this, Leibniz proceeds to state (though not, as we have said, demonstrate) the main rules of his calculus *("His positis, calculi regulae erunt tales")*:

Letting a be a given constant [quantity], da will be equal to 0, and $\overline{dax}$ will be equal to adx. If y is equal to v (that is, every ordinate of the curve YY is equal to the corresponding ordinate of the curve VV), dy will be equal to dv.

For *Addition and Subtraction:* if z − y + w + x is equal to v, [then] $d\overline{z - y + w + x}$, or dv, will be equal to dz − dy + dw + dx.

Multiplication: $d\overline{xv}$ is equal to xdv + vdx, that is, in positing y as equal to xv one obtains dy as equal to xdv + vdx. For one is entirely free to use either the expression xv or, in its place, as an abbreviation, a letter such as y. Note that in this calculus x and dx are treated in the same way as y and dy, or any other indeterminate letter and its differential.[5] Note too that the inverse procedure, based on the differential equation, is not always possible, unless [one takes] certain precaution[s] of which we shall speak elsewhere.

Next, *Division:* $d\frac{v}{y}$ or (setting z equal to $\frac{v}{y}$) dz is equal to $\frac{\pm vdy \mp ydv}{yy}$.[6]

Then, after having noted a certain number of qualifications relating to the manipulation of signs, depending on whether the ordinates increase or decrease, Leibniz embarks upon an analysis of the local behavior of curves. This question, involving the definition of convexity, concavity, and points of inflection, leads him of course to introduce second-order differentials, which he calls *"differentiae differentiarum"* ("differences of differences")"

> [. . .] if, when the ordinates v increase, the same is true of their increments or differences dv (that is, taking the differences dv as positive, the differences of [these] differences, ddv, are also positive, and similarly, taking [dv] as negative, negative), the curve turns its *convexity* towards the axis, [and] in the contrary case its *concavity*. But where the increment is maximum or minimum, that is when decreasing increments become increasing, or vice versa, there is a *point of inflection* [*punctum flexus contrarii*]; concavity and convexity turn into each other, on the condition however that in this point decreasing ordinates do not become increasing, or the other way round, for then the concavity or the convexity would not change; but it is excluded that the increments continue to grow or to diminish, when the increasing ordinates become decreasing, and vice versa. This is why there is a point of inflection when neither v nor dv are equal to 0, and that [moreover] this is the case for ddc.[7]

After making several additional remarks regarding the use of signs, Leibniz introduces two new rules, one relating to powers and the other to roots:

Powers: $dx^a = a \cdot x^{a-1} dx$, for example $dx^3 = 3x^2 dx$; $d\dfrac{1}{x^a} = -\dfrac{a \, dx}{x^{a+1}}$, for example if $w = \dfrac{1}{x^3}$ we will have $dw = -\dfrac{3dx}{x^4}$.

Roots: $d \sqrt[b]{x^a} = \dfrac{a}{b} dx \sqrt[b]{x^{a-b}}$, whence $d \sqrt[2]{y} = \dfrac{dy}{2\sqrt[2]{y}}$, since in fact in this case, a equals 1, b equals 2; therefore $\dfrac{a}{b} \sqrt[b]{x^{a-b}}$ is $\dfrac{1}{2} \sqrt[2]{y^{-1}}$;

but y^{-1} is the same thing as $\dfrac{1}{y}$, by the nature of the exponents of a geometric progression, and $\sqrt[2]{\dfrac{1}{y}}$ is $\dfrac{1}{\sqrt[2]{y}}$, $d\dfrac{1}{\sqrt[b]{x^a}} = \dfrac{-a \, dx}{b\sqrt[b]{x^{a+b}}}$.[8]

Leibniz next emphasizes the very great generality of his new algorithm, and, finally, its superiority compared with other methods then in use, whether the ones devised by Jan Hudde (1628–1704), Pierre Fermat (1601–1665), James Gregory (1638–1675), Isaac Barrow (1630–1677), or even that of John Wallis (1616–1703):

> When one knows the *algorithm,* if I may [name it thus], of this
> calculus that I call *differential,* one can find by ordinary calcula-
> tion all the other differential equations, the maxima and minima
> as well as tangents without having to eliminate fractions or other
> particularities, which, however, [is what] needed to be done with
> the methods available until the present day.[9]

Then, after having briefly presented a fresh statement of his concept of differences in relation to infinitesimal magnitudes,[10] Leibniz indicates in accordance with which procedures and manipulations it is possible to deduce from a given equation a differential
equation:

> It results from this that one can write the differential equation
> of any given equation simply by replacing each *member* (that is
> to say, each part which is only added or taken away to form the
> equation)[11] by its differential quantity. For each one of the other
> quantities (which are not themselves members but which to-
> gether help to form one), one introduces its differential quantity
> to obtain the differential quantity of the member itself, not by a
> simple substitution, but in following the algorithm which I have
> given above.[12]

The last part of the paper is devoted to the study of a few selected
examples.[13] One of them, taken from optics, was several years later
to influence Jean Bernoulli in his investigation of the brachistochrone problem.[14]

Two years after the publication of the first paper describing the
new differential calculus, in June 1686, Leibniz published (again
in the *Acta Eruditorum*) a second paper entitled "De geometria
recondita et analysi indivisibilium atque infinitorum" ("On Ab-
struse Geometry and the Analysis of Indivisibles and Infinities)."[15]
This work, whose composition was stimulated in part by his reading of a treatise on quadratures by John Craig (?1660–1731),[16]
which made use of the differential notation proposed by Leibniz
in 1684, mainly treated the inverse problem of tangents and the
calculation of quadratures in light of the new method. For this
purpose Leibniz introduced for the first time the sign ∫ and defined
the operations of "summation" and differentiation in terms of
each other. The word "integral," however, was used for the first
time in a published work only in 1690 by Jacques Bernoulli.[17]

In the 1686 paper we read:

Besides, to treat transcendental problems by the calculus,[18] when questions of areas or tangents present themselves, one can hardly imagine anything more fertile, more succinct, or more universal than my *differential Calculus or Analysis of indivisibles and infinites,* whose method of tangents, published in the Acts of the month of October 1684, gives so to speak only a brief sketch or Corollary [. . .]. I think that [Craig] himself and others will be grateful to me for giving here additional information on a subject whose utility is manifestly so important. In fact, all the theorems and problems of this sort, which were the object of a justly merited admiration, follow from it with such ease that henceforth it is no longer necessary to learn and remember them, as is the case for most of the theorems of ordinary geometry for [someone] who knows the Specious [art].[19] In the problem cited,[20] I proceed in the following way: let x be the ordinate, y the abscissa, p the distance that I have defined between the perpendicular and the ordinate; my method shows at once that one obtains pdy = xdx, which M. Craig has also shown; then, this differential equation being changed into [one of] summation, one obtains $\int pdy = \int xdx$. But in accordance with what I have shown in my method of tangents, one sees that $d\frac{1}{2}xx = xdx$; and therefore inversely $\frac{1}{2}xx = \int xdx$ (in fact, like the powers and roots of the ordinary calculus, sums and differences in my calculus, that is, $\int$ and d, are inverse). We therefore have $\int pdy = \frac{1}{2}xx$, which is what was to be demonstrated. Besides, I prefer to use signs similar to dx rather than to substitute [other] letters for them, because this dx is a certain modification of x.[21]

Due to the novelty of Leibniz's method and the obscure fashion in which he presented it, his calculus was to become known only slowly during the late 1680s and early 1690s. The first to adopt Leibnizian ideas and put them to use was the Swiss mathematician Jacques Bernoulli (1654–1705), who in May 1690 published an article employing the new calculus in the *Acta Eruditorum.*[22] The object of this article was to resolve the problem of the isochronic curve, posed by Leibniz in the *Nouvelles de la République des Lettres* in September 1687, at the height of his quarrel with the Cartesians over the "measure of forces."[23]

It was at about this time that Jean Bernoulli (1667–1748), with the help of his elder brother, was learning the new calculus. Together, in the early 1690s, the two brothers began to develop its possibilities in a systematic way,[24] largely in connection with the solving of a number of problems that scholars throughout Europe set for each other, usually in the form of a challenge. It was during a visit to Paris over the winter of 1691–1692 that Jean Bernoulli famously initiated the Marquis Guillaume de l'Hospital (1661–1704) in the arcana of the new calculus[25]—famously, since these "private paid lessons" were to provide the framework (and, to a large extent, the text as well) for the first treatise on the differential calculus, published in Paris in 1696 by the Marquis de l'Hospital under the title *Analyse des infiniment petits pour l'intelligence des lignes courbes.*[26] In this work, as was the case with Leibniz's paper of 1684, the calculus was presented as a new method for solving specific geometric problems (of minima, maxima, tangents, and so on). The use of these new procedures in physics, and more particularly in connection with the science of motion, was not immediate and required a major effort of conceptual clarification. Seeing this effort through to a successful conclusion was to be one of the chief tasks of mathematical physics in the first decades of the eighteenth century.[27]

As a result of Jean Bernoulli's *cours particuliers,* the French mathematical community, represented in this instance by a group of scholars associated with Nicolas Malebranche (1638–1715), rapidly assimilated the new Leibnizian calculus.[28] This group thus became, in André Robinet's phrase, "the sponsor of the differential calculus in France"[29] and, in particular, responsible for introducing it to the Royal Academy of Sciences.[30]

The Royal Academy, which had been founded only thirty years previously, in 1666, as part of Colbert's larger ambitions for cultural policy, was to serve as the forum for a substantial refinement of Leibnizian ideas, as well as their increasing dissemination, principally as a consequence of Varignon's vigorous initiative. The Royal Academy was also the place, it should not be forgotten, where a faction quite virulently opposed to the new calculus rose up and held forth. This opposition, extremely active from 1700 on, was led by Michel Rolle (1652–1719).[31] The burden of his critique rested on two arguments, one stressing the inadequacy and the lack of logical rigor of the fundamental concepts and principles of the new calculus, the other pretending to show (with the aid of very cleverly selected examples) that the new calculus led to error, insofar as it did not yield the same results obtained in using the classical, algebraically inspired methods of Fermat[32] and, more especially, Hudde.[33]

With regard to the first aspect of his critique, Rolle called attention to three difficulties bearing upon:

1. Higher-order differentials. In this connection Rolle was led to adapt for his own purposes certain criticisms advanced by Bernard Nieuwentijt (1654–1718) between 1694 and 1696.[34]

2. The legitimacy of neglecting the infinitesimal increase of a variable over another one equal to it. This difficulty was directly tied to one of the cornerstones of the practice of the new calculus, as this was laid down in the first "Demand or Supposition" of l'Hospital's *Analyse:*

> Grant that two Quantities, whose Difference is an infinitely small Quantity may be taken (or used) indifferently for each other: or (which is the same thing) that a Quantity, which is increased or decreas'd only by an infinitely small Quantity, may be consider'd as remaining the same.[35]

3. The legitimacy of considering differentials as absolute zeroes.

Resolving these problems proved to be an extremely delicate business that was to occupy the attention of mathematicians for more than a century. In this respect it is significant that as late as the mid-1730s one could find, for example, George Berkeley (1685–1753) repeating Rolle's arguments in support of his own apologetics in *The Analyst* (1734).[36] Berkeley subjected both the Leibnizian calculus and Newton's method of "fluxions" to harsh criticism:

> And it seems still more difficult to conceive the abstracted velocities of such nascent imperfect entities. But the velocities of the velocities, the second, third, fourth, and fifth velocities, &c., exceed, if I mistake not, all human understanding. The further the mind analyseth and pursueth these fugitive ideas the more it is lost and bewildered; the objects, at first fleeting and minute, soon vanishing out of sight. Certainly in any sense, a second or third fluxion seems an obscure mystery. The incipient celerity of an incipient celerity, the nascent augment of a nascent augment, *i.e.* of a thing which hath no magnitude: take it in what light you please, the clear conception of it will, if I mistake not, be found impossible; whether it be so or no I appeal to the trial of every thinking reader. And if a second fluxion be inconceivable, what are we to think of third, fourth, fifth fluxions, and so on without end?[37]

And, a bit further on:

> As there are first, second, third, fourth, fifth, &c. fluxions, so
> there are differences, first, second, third, fourth, &c., in an infi-
> nite progression towards nothing, which you still approach and
> never arrive at.[38]

Then, finally:

> It must indeed be acknowledged the modern mathematicians
> do not consider these points as mysteries, but as clearly conceived
> and mastered by their comprehensive minds. They scruple not to
> say that by the help of these new analytics they can penetrate
> into infinity it self: that they can even extend their views beyond
> infinity: that their art comprehends not only infinite, but infinite
> of infinite (as they express it), or an infinity of infinites.[39]

In these passages one may appreciate the extent to which what
Berkeley found new and troubling in the *differential calculus*—for
it was indeed a question of a calculus or method of calculation,
that is, a succession of operations carried out with the help of a
well-defined symbolism—were essentially the conceptual conse-
quences of the manipulation of algorithms. But Berkeley also stig-
matized the use in mathematical calculation of the first "Demand
or Supposition" in de l'Hospital's *Analyse des infiniment petits,*
which he characterized thus:

> And in general it is supposed that no quantity is bigger or lesser
> for the addition or subduction of its infinitesimal: and that conse-
> quently no error can arise from such rejection of infinitesimals.[40]

As for the third difficulty raised by Rolle—namely, "whether differentials are absolute zeroes,"—Berkeley made it one of the principal motivations for his attack in underscoring the ambiguous mode
of existence attributed to infinitesimals. We read, for example, in
The Analyst:

> I admit that the signs may be made to denote either any thing or
> nothing: and consequently that in the original notation $x + o$, o
> might have signified either an increment or nothing. But then
> which of these soever you make it signify, you must argue consis-
> tently with such its signification, and not proceed upon a double
> meaning: Which to do were a manifest sophism.[41]

This line of argument likewise followed Rolle's critique of the
conceptual foundations of the differential calculus, which aimed

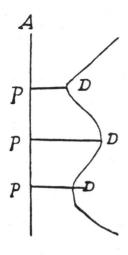

to show, relying on very cleverly chosen examples, that the new calculus led to error. A single example, taken from among the many submitted by Rolle to the Academy in 1700–1701, permits us both to grasp the nature of his approach and to better appreciate the quality of the replies given by Varignon, assisted on occasion by Jean Bernoulli.

During the Academy's sessions of 12 and 16 March 1701,[42] Rolle invited his listeners to consider "the geometric curve DD, which is formed by means of the equality B along an axis AP, as in the first figure":[43]

For the equality B we have:

$$y - b = \frac{(xx - 2ax + aa - bb)^{\frac{2}{3}}}{a^{\frac{1}{3}}}$$

To calculate the maxima and minima of this curve, Rolle proposed the application of the new method for treating infinitesimally small quantities:

To have the values of x which give the greatest and smallest [ordinates] applied to this curve, the method claims that [if] one takes the difference of the proposed equality, and if one follows the rules of analysis of infinites to take this difference, one will find it [to be] just as one finds it in C [below].[44]

$$dy = \frac{4xdx - 4adx}{3\sqrt[3]{axx - 2aax + a^3 - abb}}$$

Then he adds:

This being laid down, the method claims that the value of dy [will] be equal to zero, and [it] supposes that in [using] this method this will always result if one destroys the numerator.[45]

In writing, then, that the numerator E of the expression C is null, Rolle obtains x = a. Here the infinitesimal method seems to give only one solution, whereas "if one applies to it [to the equality B] Mr. Hudde's method, one will find these three solutions, x = a + b, x = a, x = a − b, which are distinct from each other and which are readily recognized in the first figure."[46] Since the infinitesimal method seems not to give all the maxima and minima that Hudde's method gives, "from that one can therefore be assured that this geometry leads to error."[47]

Varignon's response of 9 July 1701[48] is altogether remarkable in that it showed, on the one hand, that the application of differential methods was not done correctly and, on the other, that there was no contradiction between the new method and that of Hudde.

With regard to the first point, bearing upon the application of the differential calculus, Varignon showed (in Montucla's words) that Rolle "does not take the rule of the differential calculus in its entirety."[49] The curve in question indeed possesses a maximum (at a point of horizontal tangency) yielded by setting $dy = 0$, but it also possesses two minima (at points of vertical tangency) yielded by setting $dx = 0$ or $dy = \infty$:

> One will obtain, first, $x = a$ by setting $dy = 0$; and, second, $x = a - b$ or $x = a + b$ by making dy infinite in relation to dx, or $dx = 0$. Whence one sees that if the desired curve has some *maximum* or *minimum* that meets a tangent parallel to the axis AP, this can only be at the extremity of $x = a$; and that if it has one that meets [lines] touching [the curve (i.e., tangents)] which merge with each other, that is to say (hyp.) [tangents] perpendicular to this axis, this can only be at the extremity of $x = a - b$, and of $x = a + b$. Such that in taking A as the origin of the [abscissas] x, [the distance] A (P) = a, and, on both sides [of (P)], (P) P = b; these three values of x successively substituted in the equality B give $(P)(D) = b + b\sqrt[3]{\dfrac{b}{a}}$ as a *maximum* that terminates at a tangent parallel to the axis and (on both sides) PD = b as two *minima* meeting the tangents in D, D.[50]

This analysis then permitted Varignon to correct the shape of the curve given by Rolle, thus:

Which gives the desired curve the shape that one sees in it here [. . .] as far as the inflection points δ,δ; instead [of the shape] that Mr. Rolle believes it [to have], as in Fig. 1.[51]

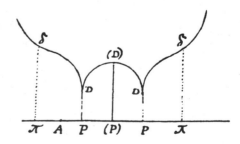

As for the second point, Varignon showed that there

was no contradiction between the results yielded by the new calculus and by that of Hudde: the fault was Rolle's for incompletely applying Hudde's method.

In modern terms,[52] this method allowed parallel tangents to be obtained along either the x-axis or the y-axis.

In the case of the tangents parallel to the x-axis:

$$F(x) = a(y - b)^3 = (x^2 - 2ax + a^2 - b^2)^2$$

$$F'(x) = x^3 - 3ax^2 + 3ax^2 - b^2x - a^3 + ab^2$$

$$F'(x) = 0 \text{ if } x = a, x = a + b, x = a - b.$$

We do indeed obtain, then, the three roots given by Rolle. In the case of the tangents parallel to the y-axis:

$$G(y) = (y - b)^3 = y^3 - 3by^2 + 3b^2y - b^3$$

$$G'(y) = y^2 - 2by + b^2$$

$$G'(y) = 0 \text{ if } y = b \text{ (double root).}$$

Now, if $y = b$, then $a + b$ and $a - b$ are roots of $x^2 - 2ax + a^2 - b^2 = 0$; and consequently, to quote the terms used by Varignon in his letter of 24 March 1701 to Jean Bernoulli:

> Therefore, of the three roots of $x^3 - 3axx + 3aax - a^3 - bbx + abb = 0$, only $x - a = 0$ gives a *maximum* ended in a [tangent] parallel to the axis; and the two others, $x - a - b = 0$, $x - a + b = 0$, alone [give] *minima* merged with tangents perpendicular to the axis.[53]

Thus, not only did the new calculus in fact supply all the desired *extrema,* it also permitted a better understanding of the properties of the curve.

As a consequence, the advent of the new technique utterly transformed the way in which mathematics was conceived, despite the weakness of its foundations. This transformation at the turn of the eighteenth century affected both the science of motion and its relation to the question of infinity.

2. The New Algorithmic Science of Motion

Since I have analyzed the emergence of the new algorithmic science of motion at length, and in great detail, in a previous work,[54] I will state the main facts of the matter here in only a cursory way. The new science involved the construction by Pierre Varignon of two related concepts—speed at each instant and accelerating force at each instant—without which it could not have been born. Their introduction dramatically altered the conceptual landscape of mathematical physics, so that questions related to the science of motion could now be reduced to problems of calculation subject to strict rules, in the present instance to simple analytical procedures governing differentiations and integrations.

2.1 Varignon's Apprenticeship in the Differential Calculus

In a letter to Jean Bernoulli dated 21 January 1693, Varignon announced that he had just "come across [. . .] the differential calculus":

> I have just come across the differential calculus as well: [though] I find it in fact much more expeditious than the ordinary [one], I do not yet clearly see the application to questions involving several differentials, differentials of differentials, etc.[55]

Less than a year later, Varignon's first known paper using the differential calculus, dated 2 January 1694, was deposited in the Pochettes de Séances of the Archives of the Royal Academy.[56] In the introduction to this paper ("General Demonstration of the Arithmetic of Infinites or of the Geometry of Indivisibles"), Varignon remarks:

> The immense utility of the arithmetic of infinites in geometry has rendered this manner of treating it very fashionable. Mr. Wallis has given us the Rules of it in his book entitled: *Arithmetica Infinitorum*; but neither he, nor anyone whom I know, has yet universally demonstrated them. Here is how it can be done using the calculus of Mr. Leibnitz.[57]

This introduction is followed by the demonstration of a theorem that applies an approach mixing Leibnizian concepts with ones derived from Cavalieri, centered on the notion of "sums":

> Theorem. Let the straight line AC be divided into an indefinite [number] of equal parts BB, such that all the [segments] AB form an arithmetic progression increasing from zero, or from A, until the largest [segment] AC. Departing from the points B, let there be as many parallels or ordinates BD in such powers n as one

likes from their origin A according to the ratio of such powers m as one likes yet again of the abscissas AB that answer to them; that is, such that everywhere $\overline{AC}^m \cdot \overline{AB}^m :: \overline{EC}^n \cdot \overline{DB}^n$. Let the exponents m and n express positive or negative magnitudes, whole or fractional. I say in general that the sum of the ordinates BD, that is, the entire figure ACEDA, is to the product EC × CA obtained [by multiplying] the number AC of these ordinates by the last AC, as n is to m + n, that is, as the exponent of the ordinates is to the sum of the ordinates and the abscissas.

Demonstr. Let the abscissas AB = x, the greatest AC = a, their differentials BB = dx; also let their ordinates BD = y, of which the last CE = b; the letter $\int$ signifies sum. That being done, given $\overline{AC}^m \cdot \overline{AB}^m :: \overline{EC}^n \cdot \overline{DB}^n$ one will have $a^m \cdot x^m :: b^n \cdot y^n$ and thence $ba^{-\frac{m}{n}}x^{\frac{m}{n}} = y$, or $ba^{-\frac{m}{n}}x^{\frac{m}{n}} = ydx = DBBD$. Therefore, $\int ba^{-\frac{m}{n}}x^{\frac{m}{n}}dx = \int ydx = \int DBBD = $ fig. ABDA. Now, according to the calculus of Mr. Leibnitz (Act. Erud. Oct. 1684) $x^{\frac{m}{n}}dx = d\frac{nx^{\frac{m+n}{n}}}{m+n}$ and $\int d\frac{nx^{\frac{m+n}{n}}}{m+n} = \frac{nx^{\frac{m+n}{n}}}{m+n}$. Therefore, $\frac{n}{m+n}ba^{-\frac{m}{n}}x^{\frac{m+n}{n}} = \int ba^{-\frac{m}{n}}x^{\frac{m}{n}}dx = $ fig. ABDA, but it has already been found that $y = ba^{-\frac{m}{n}}x^{\frac{m}{n}}$ or $xy = ba^{-\frac{m}{n}}x^{\frac{m+n}{n}}$. Therefore, in general, the figure ABDA $= \frac{nxy}{m+n}$; and, as a result, when x = AC = a, having also

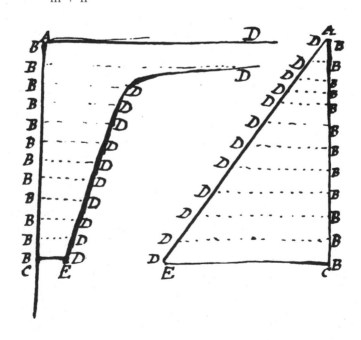

$y = EC = b$, then one will have $\dfrac{nab}{m+n}$ = fig. ACEDA, that

is, equal to the sum of the ordinates BD. But $\dfrac{nab}{m+n}$ ab

(AC × CE) :: n · m + n. Therefore, the sum of the ordinates BD is to the product of their number AC by the last CE, as n is to m + n. Which is what needed to be demonstrated.[58]

This brief paper was followed up by three others between March and December 1694, likewise deposited in the Pochettes de Séances of the Academy:

1. "Refutation of the Opinion of P[ère] Guldin and of MM. Wallis and Sturm about the Length of the Spiral of Archimedes."
2. "A Demonstration of Six Different Ways of Finding the Radii of Evolutes, Even When the Ordinates of the Curves That They Generate Converge in Any Point Whatsoever and Consequently Also in the Case That They Are Parallel."
3. "A General Way of Finding the Tangents of Spirals of All Kinds and of as Many Revolutions as One May Wish with Their Indefinite Quadratures."

These prepared the way for Varignon's paper on the "Rectification and Quadrature of the Evolute of the Circle Described in the Fashion of Monsieur Hugens," delivered at the Saturday session of the Academy on 18 June 1695 and published in the official register of minutes, marking the first time that a paper applying the methods of the Leibnizian calculus was formally retranscribed and published as part of the official records of the Academy.[59] Varignon read a further paper at the 30 July 1695 session devoted to the problem of isochronic falls.[60]

From this time the use of the Leibnizian calculus became fairly systematic in Varignon's mathematical papers. Two, in particular, perfectly illustrate his new attitude: one dated 3 September 1695, "Indefinite Rectification and Quadrature of Cycloids with Circular Bases, Whatever Distance May Be Assumed between Their Point of Description and the Center of Their Moving Circle,"[61] and another dated 12 November 1695, "Of the Evolution of Spirals of All Kinds, in Which It Is Shown That They All Evolve into Parabolas of Solely a Degree Higher than Their Own, with a General Method for All These Sorts of Evolution."[62]

Varignon, having mastered the essential methods of the Leibnizian calculus, was now to adapt the new theoretical framework to a renewed study of motion. A new conceptualization of velocity was now possible.

2.2 Varignon's Construction of the New Concepts of the Science of Motion

In elaborating the concept of "speed in each instant" in two papers read before the Royal Academy at the Saturday sessions of 5 July and 6 September 1698, respectively, Varignon gave the science of motion a new point of departure. The paper of 5 July, "A General Rule for All Sorts of Motions of Whatever Speeds Varied at Pleasure," for the most part treated only rectilinear motions, while the paper of 6 September, "Application of the General Rule of Speeds Varied as One May Please to Motions through All Sorts of Curves, Mechanical and Geometrical Alike: Whence Is Deduced Yet Another New Way of Demonstrating Isochronous Falls in the Inverted Cycloid," was devoted to motions along curvilinear trajectories.[63]

The object of the first paper therefore was to give a general expression of speed that would permit all motions to be treated in the case of rectilinear trajectories, no matter how the speed varied:

> [. . .] here then is a [rule] which is still infinitely more [useful], in that it is suited not only to accelerations or retardations corresponding to all these powers of spaces traversed, but to all sorts of speeds varied as one may please, and to all that one may speculate about [regarding] the relation of spaces and times.[64]

Varignon's approach as a whole consisted in considering the speed of a body as uniform during each instant, or, as Fontenelle characterized it:

> [. . .] M. Varignon has not failed to treat varied motions as uniform [motions], & to draw from the one [sort] the same consequences as from the other.[65]

In his paper, Varignon first of all defines a certain number of variables that are to serve as the conceptual basis for his project.

> [. . .] (all the rectilinear angles that one sees here being right [angles]) let $AB = x$ be the distances traveled in whatever direction one pleases, $BE = z$ the times spent in traveling through them, and $BC = y = DF$ the speeds at each point B of these distances.[66]

The variables for space, time, and speed now being represented by the "axes," or line segments AB, BE, and BC (or DF), it then becomes possible, by associating these variables in pairs, to generate what might in modern terms be called the graph of the spaces EE and the two graphs of the speeds CC and FF:[67]

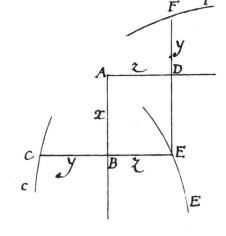

[. . .] such that the curve EE (any whichever) formed by all the points E, expresses the spaces [traversed] by its abscissas AB and the time spent traversing them by its corresponding ordinates BE; the curve CC (again, any one whichever) formed by all the points C, likewise expresses the spaces by the abscissas AB and the speeds at each point B of these spaces by its corresponding ordinates BC; finally, the curve FF (any one whichever, as before) also expresses by its ordinates DF these same spaces compared to the times BE or AD that serve as their abscissas.[68]

Let us now consider what happens in an instant:

This being laid down, the instants will be $= dz$;[69] the space traversed in each instant will be $= dx$; and the speed with which dx will have been traversed will be $= y$.[70]

Thus, in an instant equal to dz, the distance traveled by a moving object will be equal to dx, and the speed of this object will be equal to y. Varignon asserts then that this speed, in each instant, can "be regarded as uniform":

Such that this speed (y), in each instant being able to be regarded as uniform, because $y \pm dy = y$, [and so] the notion of uniform speeds by itself will give $y = \dfrac{dx}{dz}$ as the rule of [as many] varied motions as one pleases, that is to say, whatever relation of space, time, or speed that one may suppose; the speed of each instant being always and everywhere equal to the quotient of the space traversed in each instant divided by that same differential of time.[71]

It is important to emphasize, as Varignon himself would do (though not explicitly until nine years later, in a paper dated 6 July 1707, "Of Motions Varied at Will, Compared among Themselves and with Uniform Ones") that "space and time being heterogeneous, they are not strictly the ones being compared together in the relation that is called speed, but only the homogeneous magnitudes that express them, which here are, and will always be in what follows, two lines, or two numbers, or any two other such homogeneous magnitudes as one may wish."[72] This passage, while it does put forth the first elements of a genuine theory of measurement, nonetheless remains highly unsatisfactory. In particular, the absence of any reference to magnitudes of the same kind considered as units immediately introduces an imprecision regarding what properly would be called the definition of instantaneous speed. For this reason, but also because Varignon's concept of speed in each instant is not actually conceived as a limit toward which a certain quotient tends but only as a quotient, we will not refer in what follows to a definition of the concept of instantaneous speed, although Varignon sometimes used this expression,[73] nor even to a definition of the concept of speed in each instant, but more modestly to a characterization of the concept of speed in each instant.[74]

This being the case, it is clear that the reasoning which led Varignon to the concept of speed in each instant rests directly on the methods of the calculus of differences, particularly as they were employed in de l'Hospital's *Analyse des infiniment petits* of 1696. The first "Demand or Supposition" of this work, it will be recalled, was as follows:

> Grant that two Quantities, whose Difference is an infinitely small Quantity may be taken (or used) indifferently for each other: or (which is the same thing) that a Quantity, which is increased or decreas'd only by an infinitely small Quantity, may be consider'd as remaining the same.

Varignon was therefore justified in treating y ± dy as equivalent to y. This speed y could thus be "consider'd as remaining the same" during the interval of time dt(dz) in the course of which the distance dx is traversed. And, as a result, "the notion of uniform speeds by itself" did in fact immediately yield the expression of the speed y in each instant y = dx/dt. Varignon made the sense of the concept of speed in each instant (or, as he put it, of instantaneous speed) explicit in the posthumously published treatise on the motion and measurement of running and flowing waters (1725):

> Note [. . .] however each instantaneous speed is equal and uniform in itself, because in supposing that it corresponds to an infinitely small instant, it can undergo during this infinitely small instant only an infinitely small variation, [which] as a result [is] null by comparison with the variation which occurs in a finite time.[75]

Varignon's characterization of speed in each instant permitted him to state a "general Rule" in his paper of 5 July 1698.[76] This was really only a restatement of the initial characterization, in three different ways, privileging in turn speeds, times, and spaces, thanks to the power of algebraic manipulation:

General Rule

of speeds	of times	of spaces
$y = \dfrac{dx}{dz}$	$dz = \dfrac{dx}{y}$	$dx = ydz.$

Of these three formulas, it follows quite obviously:

> Whatever presently may be *the speed of a body* (accelerated, retarded, in a word varied as one pleases), *the space traversed,* and *the time spent in traversing it;* two of these three things being given at pleasure, it will always be easy to find the third by means of this rule, even in [the case of] the most bizarre variations of speed which can be imagined.[77]

Varignon's construction of the concept of speed in each instant was therefore finally made possible by a twofold conceptual innovation:

• on the one hand, by advancing to some extent the traditional debate over homogeneous magnitudes, making it possible to directly express the concept of speed in the form of a quotient;

• on the other, by introducing the idea that in an instant of time speed can be considered to be constant, making it possible to readily apply the concepts of the Leibnizian calculus.

The first part of the paper of 5 July 1698 concludes with the statement of the "general Rule." The second part attempts to apply it by means of five examples.

Two months later, in the paper of 6 September 1698, Varignon completed the work of the first by considering motion along curvilinear trajectories. At the outset of this paper he recalls the results of the earlier paper:

> 5 July last [1698] I proved in general to the Academy that with what[ever] variation of speed that a body may move, that [speed] which it has at each instant is always equal to the quotient of what space it then traverses divided by this instant.
>
> Whence I conclude that in taking AB = x for the whole space traversed, the ordinates BC = y of any curve CC for the speeds at each point B of this space, and the ordinates BE = z of any other curve EE for the times spent in going from A to B; one will always have $dz = \dfrac{dx}{y}$.[78]

It was now a question, for Varignon, of generalizing this result to the case of a motion along, not a rectilinear trajectory, but a curvilinear trajectory.

Varignon thus regarded uniform speed in each instant as equivalent, no longer to the relation of an infinitely small line segment dx to an infinitely small time dz, but to that of an infinitely brief arc of a curve (treated as a line segment whose length could be determined by applying Pythagoras's theorem in the case of infinitesimally small increases)[79] to an infinitely small time:

> [. . .] if to the place AB the body moves with the same speed corresponding to AK (parallel to the ordinates) in G along whatever curve GG[80] whose ordinates are BG = V, during times expressed by the ordinates BH = s of still any other curve whatever HH; similarly it follows that one will always obtain $ds = \dfrac{\sqrt{dx^2 + dv^2}}{y}$, or (assuming a = 1 in keeping with the law of homogeneous [quantities]) $ds = \dfrac{a}{y}\sqrt{dx^2 + dv^2}$, dz being converted here to ds, and dx to $\sqrt{dx^2 + dv^2}$.[81]

This passage makes it clear that Varignon preferred to speak of the instant of time ds rather than the speed in each instant y. As the rest of the paper shows in greater detail, what interested him here, more than discovering the speed, was discovering the time required to traverse all sorts of curves, mechanical and geometrical alike, speeds and trajectories being given. The main objective of the paper, as the title makes plain, continued to be the demonstration, by means of this new method, of the isochronic character of falls in the inverted cycloid:[82]

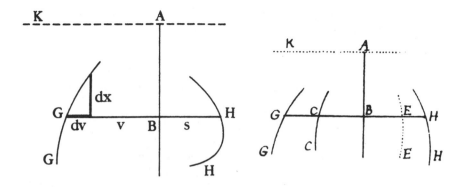

[. . .] if one substitutes [for] the value of y in x the equation of the curve CC, and again [for] v in x the equation also given of the curve GG, one will have in x and in s the desired equation of the curve HH whose ordinates BH(s) express the time spent in going from AK to G along the curve GG.[83]

Thus ends the first part of the paper, the second part being devoted, as in the preceding paper, to various examples. These examples permitted Varignon to highlight the pertinence of his new concepts to the treatment of problems of motion. It remained for him then only to make a concluding statement about the general import of the papers of 5 July and 6 September, as well as about the fruitfulness of the rules now established:

These two examples suffice to show how one is to make use of the formula $ds = \dfrac{a}{y}\sqrt{dx^2 + dv^2}$ in order to find the time spent in traversing all sorts of curves, mechanical and geometrical alike, no matter how varied the speed[s] of the bodies traversing them may be, having only to substitute any such other curve as one likes in place of the parabola whose ordinates have just been used to adjust these speeds in order to accommodate Galileo's hypothesis regarding the fall of bodies. Not only can the curve of the times thus be found, that of the paths and of the speeds being given; but each one of these [three curves] may be found, [any] two others being given, in proceeding as in the examples of the general Rule which I demonstrated last 5 July at the Academy: this is why I do not dwell on it further [here].[84]

In order to fully appreciate how important the transformation of the conceptual landscape of the science of motion brought about by the construction of the concept of speed in each instant was, it suffices to read, for example, the first pages of the paper that Varignon devoted in 1707 to the motion of projectiles in resistant media.[85]

After giving various definitions directly related to this topic (of instantaneous resistances, total resistances, primitive velocities, and so on), Varignon introduces two lemmas. The first is intended mainly to specify the sense in which the expression "infinitely small Resistances" is to be understood in the context of the study of motion in resistant media, while the second, through its use of the concept of speed in each instant, perfectly illustrates the conceptual transformation that has taken place:

> Lemma II. The sum of the instantaneous speeds of a body moved in any manner whatever is always proportional to the length of the path that they cause it to traverse [during] one instant after another.[86]

While the expressions "sum of the speeds" and "proportional to the length of the path" may recall earlier ways of conceptualizing problems of motion (see in particular chapter 3, above), the expression "one instant after another" (literally, "one [speed] after another per instant") seems by contrast to be inspired by a wholly different way of conceptualizing such problems. The situation becomes clearer several lines later, where one encounters the following "demonstration":

> Let e be this distance traveled during time t and de the distance traveled during each instant dt, with an instantaneous speed called u. This speed consisting only in the relation of de to dt, it is manifest that one will have here $u = \dfrac{de}{dt}$, or udt = de. Therefore too $\int udt = e$.[87]

The calculation of the distance traversed appears now as a trivial consequence of the characterization of the speed in each instant. From now on this speed would constitute the primary concept of the science of motion, on the basis of which the science could at last be deductively and algorithmically organized.

The actual construction of the algorithmic science of motion was carried out by Varignon some months later in two papers: the first, "A General Way of Determining Forces, Speeds, Spaces, and Times, Only One of These Four Things Being Given, in All Sorts of Rectilinear Motions Varied at Pleasure," was dated 30 January 1700;[88] the second, "Of Motion in General through All Sorts of Curves; and of the Central Forces, Centrifugal and Centripetal Alike, Necessary to the Bodies Which Describe Them," was dated 31 March 1700.[89]

The first paper was devoted to bodies which, being subjected to the action of a central force, describe rectilinear trajectories; and the second, to bodies describing curvilinear trajectories under the same conditions.

Inspired by Newton's results, particularly those contained in the third corollary of lemma X of the first section of the *Principia*, Varignon managed in both cases to express the "accelerative force" in each instant. Rather than go into great detail, having already analyzed his approach, I will limit myself here to giving the "general Rules" formulated by Varignon in each case while preserving his notation (x for distance, s for curvilinear distance, t for time, v for speed, y for accelerative force):

General Rules of Motion in Straight Lines[90]

$$1. \quad v = \frac{dx}{dt} \qquad\qquad 2. \quad y = \frac{dv}{dt}\left(\frac{ddx}{dt^2}\right)$$

General Rules of Motion in Curved Lines[91]

$$1. \quad v = \frac{ds}{dt} \qquad\qquad 2. \quad y = \frac{dsdds}{dxdt^2}\left(\frac{vdv}{dx}\right)$$

These "general Rules" make it clear that the expressions of speed in each instant and of "accelerative force" in each instant, which were *successively* constructed by Varignon, can in fact be deduced from each other by a simple calculation applying the algorithms of the Leibnizian calculus. In this sense it may be said that the problems of the science of motion now reduced, in Comte's phrase, to "simple analytical techniques that will consist either in differentiations or, more often, in integrations."[92]

Construction of algorithms, reduction of problems to simple analytical techniques, calculations of all kinds—geometrization now seemed far away indeed. What Varignon had done was to show, in an exemplary and, at long last, inaugurational way, that scientific work must aim above all at obtaining, and rigorously manipulating, *rules* and *formulas*. The field of modern mathematical physics was now entered into once and for all, and that of the old science of motion, with its ontological and geometric ambitions, was left behind.

What, then, did these new rules and formulas portend for the future? More specifically, what was the meaning of this conquering spirit of mathematization and of the novel methods made possible by this new mathematical physics?

Fontenelle and the Reasons of Infinity

In 1697, Bernard Le Bovier de Fontenelle, by that time a member of the French Academy for six years, was elected to the Royal Academy of Sciences, succeeding Jean-Baptiste Du Hamel (1624–1706).[1] Two years later, in 1699, as part of the reorganization of the Academy, he became its first permanent secretary. It was from this latter date that what amounted to the second phase of the Academy's existence began. The Academy was created in 1666, under the supervision of Colbert and the protection of Louis XIV, but its official rules and regulations were not fixed until more than thirty years later. In 1699 it ceased to be an independent institution, albeit one closely allied with the government, and became instead an institution of the state.

This restructuring of the Academy coincided with the decision to annually publish a volume containing the papers delivered at the sessions of the Academy that would contain a lengthy introductory section called "History of the Academy." The drafting of this part of each volume was confided to the permanent secretary. Eventually these volumes came to bear the general title *Histoire de l'Académie Royale des Sciences avec les Mémoires de Mathématique et de Physique pour la même année.* The first volume of the series, recording the events of the year 1699, was published in 1702 and the last volume, corresponding to the year 1790, was published in 1797.

In addition to editing the introductory sections of these volumes, Fontenelle was assigned responsibility for writing the history of the first period of the Academy's existence, covering the years 1666–1699. This labor resulted in the publication in 1733 of two volumes: *Histoire de l'Académie Royale des Sciences depuis son établissement en 1666 jusqu'à 1686* and *Histoire de l'Académie*

Royale des Sciences depuis 1686 jusqu'à son renouvellement en 1699. Fontenelle actually wrote only the part concerning the period 1666–1679, basing his account on Du Hamel's *Regiae Scientiarum Academiae Historia* (Paris, 1698). The foreword to the first volume of the *Histoire* notes:

> The History of the Royal Academy of Sciences that we publish today, has been made in part on [the basis of] the Records of this Company, and in part on the Latin History of M. Du Hamel. M. de Fontenelle, permanent Secretary of the Academy, has taken this History from the origin of the Academy until almost the end of the year 1679. The other years until 1699, when the great series of the History and the Papers begins, have been put in French, roughly in accordance with the order observed by M. de Fontenelle in the preceding [years]: in both [volumes] will be found things which were omitted by M. Du Hamel; and, oppositely, M. Du Hamel has included items that will not be found here, either because they have been wholly restated or else subsequently treated more broadly by the members of the Academy, or because the first Years not being absolutely similar to those of the Latin History, it was not believed that the following ones had to be [made] more conformable to them.

Fontenelle's editorial activity in his capacity as permanent secretary of the Royal Academy until 1740, when he was succeeded by Jean-Jacques Dortous de Mairan (1678–1771), was therefore very considerable.[2] This is all the more important since, in the sections of the annual volumes devoted to the history of the Academy, Fontenelle attempted to place in perspective the various papers published in these volumes, with the aim of clarifying the historical and scientific context in which new research was being conducted. This analysis, often quite detailed, and calling attention to points of the greatest importance, particularly ones of a scientific nature, was accompanied by a number of more or less cursory remarks that would unquestionably be regarded today as epistemological in character.

Throughout his work Fontenelle inquired into the nature and aims of mathematics and scientific method, the relation between mathematics and physics, the role of basic research, and the uses of science and technology. This amounted to a profound analysis of fundamental questions, based on science as it was actually practiced. Fontenelle's reflections, frequently limited in the "historical parts" to brief remarks, suggestions, and expressions of intuitive insight, were to assume a much more systematic and developed form, at least with regard to the set of questions pertaining to the status of mathematical objects, in his *Élemens de la géométrie de l'infini*, published in 1727. In the various writings that make up this unjustly neglected work, Fontenelle elaborated one of the first and most profound meditations on the meaning of mathematical physics and on the requirement of total mathematicity that was to govern the further development of this branch of knowledge.

Mathematics played a decisive role in the development of scientific knowledge at the beginning of the eighteenth century. The rigor of its argumentation (whether, for example, in the case of central forces or of the motion of projectiles in resistant media) depended to a large extent, as we have already seen, on the new mathematics itself, on its internal organization and coherence. The considerable development of a mathematized science of nature was in turn to a great degree the result of Leibniz's recent advances. But though the enormous usefulness of this new mathematics was quite readily acknowledged at the time, it was far from clear that its foundations were secure and its methods justified. One major task remained to be accomplished: to transform a fertile method—an art—into a genuine science. This task was inevitably accompanied by another: to carry on Varignon's work in algorithmizing the science of motion, which meant deliberately and self-consciously placing it within the conceptual framework of mathematical physics.

Not the least of Fontenelle's contributions were his consideration of problems too often evaded by his contemporaries, concerning the foundations of the new Leibnizian calculus and the status of the mathematical style of physics associated with it, and his attempt to supply answers to these problems, both in his *Elements* and in his many critiques of papers by academic colleagues, as well as in the funeral orations that for more than forty years he composed to mark their passing.

1. The Mathematics of Infinity

In December 1727, in Paris, Fontenelle brought out his *Elements,* a work that was dear to his heart and to which he had devoted nearly thirty years of work. In its original quarto edition, printed at the Royal Press and containing 548 pages divided into two parts,[3] it was presented to the public as a "series of papers of the Royal Academy of Sciences."[4] The book was accorded a distinctly reserved reception by Fontenelle's contemporaries. Rather than pause to reflect upon the larger intellectual project represented by his work, they were intent solely on calling attention (though often rightly) to its mathematical inadequacies and certain flaws in theory and design. As a result, they failed to recognize what was really at issue and what Fontenelle was trying to do.

As far as Fontenelle himself was concerned, the object of the book was not to present new results but rather, in refusing to reduce the new calculus to either a simple method of approximation[5] or a mere technique of calculation,[6] to illuminate in detail such results as had already been obtained by means of what we would today call an inquiry into the foundations of mathematics.[7] Unlike many others of the period, Fontenelle believed that one should not be satisfied with a method that incontestably "worked well" and so yielded a good many results but that nonetheless one manipulated blindly, as it were. Fontenelle is very explicit on this point in the remarkable essay with which he prefaces his book and which itself in some sense constitutes a history of the genesis of the new Leibnizian calculus:

> [. . .] a bizarre thing has occurred in the high[er] Geometry: certainty has become drowned in clarity. One still holds on to the thread of the calculation, [that] infallible guide; it does not matter where one ends up, one needs [only] to arrive, whatever darkness may be found there. Moreover, glory has always been attached to grand investigations, to the solutions of difficult Problems, & not to the clarification of ideas.
>
> I have believed that this clarification, neglected by skillful Geometers, might be useful in Geometry; one will not walk ahead more surely, but one will see more clearly [in looking] around, [for in addition to] the thread that one had in the dark Labyrinths, one will have a torch whose gleam, be it ever so small, will still be of some use, & indeed if this feeble glow that I present is not false, nothing will prevent it from being greatly augmented.[8]

And further still:

> I recognize that one may reproach me [in saying] that instead of illuminating the Infinite, I have brought a new obscurity to it, [in the form of] a Paradox unheard of, which is expounded in Sect. III, & which afterward is often found throughout the Work: but if this Paradox is true, if it necessarily follows from the nature

of the Infinite, I cause [this nature to be] better known, [and] I make better known its properties, which, although obscure, are the source of all that which is most astonishing which the Calculus gives us; one will arrive at [even] greater marvels well prepared, & without that kind of surprise which is not worthy of a true Science in [its] foundation. It is always illuminating to some [small] degree, to [clearly] see to what principle, even if little known, certain effects are due.[9]

To satisfy this requirement of clarity, but also of rigor, Fontenelle proposed to construct a genuine theory, or "general system of the infinite,"[10] capable of accounting for all the results so far obtained and of giving them a meaning:

When a Science, such as Geometry, has only just been born, one can scarcely catch hold of [more] than scattered Truths that do not hang together, & one proves them one at a time as [best] one can, & almost always with much difficulty. But when a certain number of these disconnected Truths have been found, one sees how they fit together, & general principles begin to emerge, not yet the first or the most general, [for] a larger number of Truths are needed to make them appear. Several small Branches, which one holds on to separately at first, lead [one] to the great Branch that produces them, & several great Branches finally lead to the Trunk. One of the great difficulties that I experienced in the composition of this Work was grasping the Trunk, & several great Branches seemed to me to be [the Trunk] which were not. I am not sure that I am yet not mistaken, but when finally I took the Infinite for the Trunk, it was not possible for me to find another one, & I saw it distribute everywhere, & spread its boughs with a regularity & a symmetry which has served my particular cause to no small [degree].

One advantage of having grasped the first Principles would be that order would establish itself everywhere almost by itself, this order which embellishes everything, which fortifies Truths by their connection, which those to whom one speaks have the right to demand, & which one cannot refuse to them without [committing] a kind of injustice, above all if one sacrifices their convenience to the glory of appearing more profound.[11]

Or again, a bit further on:

Calculation in Geometry is scarcely [anything other] than what experiment is in Physics, & all the Truths produced solely by Calculation can be treated as Truths of experience. The Sciences must go as far as the first causes [of things], above all Geometry, where one cannot suspect, as in Physics, [the existence

of] principles which are unknown to us. For if there are in Geometry, so to speak, only [those ideas] which we have placed
there, these are only the clearest ideas that the human Mind can
form about Magnitude, compared together, & combined in an
infinity of different ways.[12]

The meaning of this theoretical undertaking, aimed at constructing a "general system of the infinite," was not always well understood. Thus, for example, Leibniz wrote to Varignon in a letter
dated 20 June 1702 as follows:

> Between us, I believe that Mons. Fontenelle, who has a fine
> and gentlemanly mind, wished to mock [us], when he said that
> he wished to make metaphysical elements of our calculus. To tell
> the truth, I myself am not overly persuaded that it is necessary to
> consider our infinites and infinitely small [quantities] otherwise
> than as idealized things or as well-founded fictions. I believe that
> there is no created [thing] below which there is not [in principle]
> an infinity of [others], however I do not believe that there are
> any [in actuality], nor indeed that there are any infinitely small
> [things], and this is what I believe [myself] able to demonstrate.
> Namely, that simple substances (that is, [substances] that are not
> entities by aggregation) are truly indivisible, but they are imma-
> terial, and are only principles of action.[13]

Père Castel, on the other hand, in a letter addressed to Fontenelle
more than twenty-five years later, on 20 March 1728, regretted
that the permanent secretary of the Royal Academy had not "returned to metaphysics" in his recently published work:

> All the others, not excepting M. de l'Hôpital, have treated only
> the art of it, the trial and error and the routine of the calculus.
> So that if you had wished to return to metaphysics, as I had al-
> ways hoped, I do not see what could be lacking to so fine a
> science.[14]

The reason for this incomprehension resided to a large degree in
the failure of both men to grasp Fontenelle's concept of geometric
system, the key to unlocking the meaning of the essential distinction made in the *Elements* between geometric infinity and metaphysical infinity. Fontenelle insisted on the importance of this concept in giving the first part of the work the title "General System
of the Infinite." Moreover, it was precisely to this question that he
called the attention of his correspondents. Thus, in his letter of 22
April 1725 to Jean Bernoulli, he "flattered" himself that his "ra-

ther bulky work, whose title is *Élemens de la géométrie de l'infini*," might be taken to be

> [. . .] a kind of *system*, not Metaphysical, but Geometrical, well enough bound together with all that which we have discovered on this great subject. The order of it I believe [is] nearly as exact as it can be, and the spectacle [it presents] rather fine for a mathematical Mind [to behold]; it was necessary, if only for binding together the stones of the Building, that I mixed a great number of thoughts which were mine alone together with those which belonged to you.[15]

But what, for Fontenelle, most perfectly illustrated his conception of a geometric system "well bound together" was the decisive place that he found himself obliged, in order to ensure the coherence of the system, to accord to a paradox: the paradox that led to the introduction of "finite indeterminables."

Thus, in the same letter to Bernoulli, he adds:

> For what is bizarre is that in the same measure as this principle is *paradoxical* and *wild*, it is general and fruitful, and I beg you on this point simply to take me at my word. I come across it everywhere, and without in the least having looked for it, quite to the contrary. I would have wished with all my heart to have been able to do without it, for I knew the danger of it. I find at each moment in the course of the work new proofs of it by analogies, by the Calculus, by the necessary connection of this principle with all the constant truths which may be related to it.[16]

Fontenelle returns repeatedly to these same themes not only in his correspondence with Jean Bernoulli[17] but also with Jean-Pierre de Crousaz (1663–1750),[18] Willem Jakob s'Gravesande (1668–1742),[19] and David Renaud Boullier (1669–1759).[20] Reading these letters, it is evident that, for Fontenelle, his *Elements* presents a "geometric system" endowed with a remarkable internal coherence ("well bound together" and relying, among other hypotheses, upon one in particular that took the form of a paradox leading to the introduction of "finite indeterminables." On this view, the existence of the objects of the system rests ultimately upon their internal coherence. It is this coherence that guarantees their reality, that underwrites their sole claim to existence. As Fontenelle expresses the point in the preface to his *Elements:*

> Geometry is wholly intellectual, independent of the actual description and existence of the Figures whose properties it dis-

covers. All that which it conceives [to be] necessary is real [by virtue] of the reality that it supposes in its object. The Infinite that it demonstrates is therefore as real as the Finite,[21] & the idea that [Geometry] has of it is not in any way different from the others, an idea of supposition, which is only convenient, & which must disappear once use has been made of it.[22]

From this Fontenelle's distinction between geometric infinity and metaphysical infinity takes on its full meaning:

> We [all] naturally have a certain idea of the Infinite, as a magnitude without bounds in any sense, which comprehends everything, beyond which there is nothing. One may call this Infinite *Metaphysical*: but the *Geometrical* Infinite, that is to say, that which Geometry considers, & of which it has need in its investigations, is quite different, [for] it is only a magnitude greater than any finite magnitude, but not greater than any magnitude [at all]. It may be seen that this definition allows Infinites smaller or greater than other Infinites, & that [the definition] of the Metaphysical Infinite does not. One is therefore not justified to draw objections from the Metaphysical Infinite against Geometry, which takes into account only that which it encloses within its own idea, & nothing of what belongs to the other.[23]

In the context of Fontenelle's "geometric system," the geometric infinite therefore appeared as a mathematical concept which, to the extent that it was mathematical, was ontologically independent of metaphysical infinity. It arose only from the coherence of the system within which it was deployed. As a result, for Fontenelle, no critique of the concept of geometrical infinity relying on the concept of metaphysical infinity—a rather vague notion in his view in any case—had any value whatsoever.[24] By virtue of its willingness to consider geometric infinity as a specific concept in its own right, the content of which needed to be defined in terms of mathematical language alone, Fontenelle's work (despite certain mathematical weaknesses, to which we shall return, resulting for the most part from the lack of a clear distinction between ordinal and cardinal numbers) incontestably prefigured that of Cantor and his successors.[25]

In the first section of his *Elements* Fontenelle defines magnitude as that which "is susceptible of augmentation and diminution, or, what is the same, of [being made] more or less." Magnitudes will therefore be "numbers, lines, surfaces, solids, times, etc."[26] In its general sense, magnitude is therefore always "by its essence, sus-

ceptible of more or less," as a result of which "it loses nothing of
its essence in receiving this more or less, [and] therefore it is still
magnitude, therefore still equally susceptible of more or of less,
therefore it is always susceptible of them; therefore it is without
end, or at the infinite."[27] The object of the second section is to
examine just this "infinitely great magnitude."[28]

The particular reality of an "infinite number," as Léon Brunschvicg noted in connection with Fontenelle, is "immediately given"[29]
by the "natural series of numbers whose origin is 0 or 1," and, in
this sense, an "infinite number" possesses the same type of reality
as that which one grants to finite numbers:[30]

> 84. To better conceive the Infinite, I consider the natural Series
> of numbers, whose origin is 0 or 1.
>
> Each term always increases by a unit, & I see that this aug-
> mentation is without end, and that however great may be the
> number [at which] I arrive, I am not nearer to the end of the
> series, which is a characteristic that cannot suit a Series the num-
> ber of whose terms is finite. Therefore the natural Series has an
> infinite number of terms.
>
> It would be futile to say that the number of terms which com-
> poses it is actually always finite, though I can always augment it.
> It is indeed true that the number of terms which I can actually
> run through or arrange according to their order is always finite;
> but the number of terms of which the Series itself is composed is
> another matter. The terms of which it is itself composed all exist
> equally, & if I conceive it [to be] carried out to 100, I do not
> [thereby] give to these 100 terms an existence of which all the
> others that lie beyond [them] are deprived. Therefore, all the
> terms of the Series, though they may not all be taken in or con-
> sidered together by my mind, are equally real. But [since] the
> number of them is infinite, as has just been proved, therefore an
> infinite number exists just as truly as [do] the finite numbers.[31]

Additionally, "in the natural series each term is equal to the
number of terms which [run] from 1 up to [and] including it"; but
since "the number of all its terms is infinite," it follows as a result
that the natural series "has a last term which is this very infinite."
This last term is expressed by the "character ∞."[32]

However, although passing from the finite to the infinite in the
natural series of numbers may be "inconceivable," this situation
did not in the least hinder a mathematical analysis of the infinite
since the infinitely great magnitude, according to Fontenelle, must
be taken "not as being in this obscure passage from the finite to

the infinite; but as having jumped over it entirely and having passed through [all] the necessary degrees, whichever they may be, [even] if I may be able only sometimes to [cast] some light on the nature of these degrees." [33]

Just the same, the very concept of infinitely great magnitude seems contradictory since, on the one hand, "the natural idea of infinite magnitude is that it may not be greater or augmented," and, on the other, infinitely great magnitude insofar as it is magnitude "must preserve the essence of it and be capable of augmentation and indeed without end." [34] Yet "these two ideas so contrary in appearance are perfectly reconciled" if one rigorously distinguishes, following Fontenelle, the order of the finite from that of the infinite. The force of this is to say that the infinite, insofar as it is magnitude, may receive augmentations and diminutions, but only augmentations and diminutions that belong to its order. As a consequence:

> 113. These orders thus established, any magnitude whatever has finite relations with all that which is of its order, & it can only receive augmentations and diminutions by [the addition or subtraction] of what is of its order. If one conceives it [as being] raised to a higher order, it is necessary to take it as having jumped over this immense passage, & then all that which is of a lower order is no longer magnitude by comparison with it, & disappears before it, [just] as [it] is not itself magnitude by comparison with all those of higher orders, & would disappear before them. All this is only what has been said of the Finite and of the simple Infinite, applied to all the orders in general, of which the Finite and the simple Infinite are only the first two. [35]

Fontenelle next brings out two key concepts—the "indeterminate infinite" and the "indeterminable finite"—which serve as the cornerstone of his system to the extent that they are introduced implicitly to avoid both the objections brought against the concept of actual infinity and a return to Zeno's paradoxes.

The introduction of the concept of the "indeterminable finite" assumes particular importance in this case, since, as we shall see, it was intended as a direct response to the formulation of a paradox that could be used as an argument against actual infinity: [36]

Indeterminate infinites. By dividing the last term of the series of natural numbers A (i.e., ∞) by "any finite number" n, one obtains infinites of the same order:

Since A is an arithmetic progression, the last term of which is ∞, its middle term is $\frac{\infty}{2}$, Infinite, after which there can only be larger Infinites. Similarly, the term of its 1st quarter is $\frac{\infty}{4}$, again infinite, that of its first 100th part is $\frac{\infty}{100}$, infinite as well; such that of the infinite interval, which is between 1 and ∞, divided into 100 parts, already [the last] 99 of them can have only infinite terms, & only the first remains which may have finite [ones]. It can be seen that this first 100th part will be infinite, since it will be a finite part of an infinite interval, & consequently it will yet contain an infinity of terms.[37]

These infinites, contained in A in "prodigious numbers" and "less than ∞," are all designated "by this character $\propto$, which represents an indeterminate and variable infinite of the same order as ∞, which is a fixed infinite."[38] As a result, these indeterminates are such, for example, that "$\frac{\infty}{\propto}$ is a finite integer, plus a fraction in most cases, and $\frac{\propto}{\infty}$ is a finite fraction less than 1," or again, since the indeterminate infinites have only finite relations with each other, "$\frac{\propto}{\propto}$ is a finite."[39]

These results are very important from the perspective of Fontenelle's system, for they enable the relations of infinite magnitudes of the same order to be expressed in finite terms.

Indeterminable finites. Fontenelle stresses that "it can be seen that A^2 has as many terms as A";[40] however, it must be admitted that A^2 has infinites "sooner than A"—that the passage to the infinite occurs sooner in A^2 than in A because the terms of A^2 increase as the squares of A:

I present the two Series to view, the better to display their corespondence:

A	1	2	3	4	&c.	n	nn	B	infinites	∞
A^2	1	4	9	16	&c.	nn		C	infinites	∞^2

The line BC marks in A the separation of the finite terms from the Infinites, such that at the left of BC they are all Finite, & to its right Infinite, & and at the same time it marks in A^2 that, at least to its right, they will all be Infinite, for the Infinites of A can only augment in A^2 by being raised to squares.

Let nn be the largest finite square that [there] is in A, & consequently written down to the left of BC, & just next [to it]: it will also be in A^2, since it is the square of n, one of the terms of A. But it will be in A^2 under n, its root, & n is in A, far removed from nn, & all the further as n is larger. But nn is the largest possible finite square, & in A^2 there is still [further to go] from nn to the line BC. Therefore, in A^2 there are no more finite terms after nn, and indeed, in this Series there is a void from nn until the line BC; such that all the Finite terms which are in A from n until the line BC have no correspondents or squares in A^2, which manifestly is impossible. Therefore, after nn there come Infinites in A^2, & sooner [in] A^2 than [in] A.[41]

Consequently, and paradoxically, as Fontenelle points out, finite terms in A can yield infinite squares:

[. . .] the infinites which will be in A^2 from nn until the line BC will therefore be squares of corresponding finite terms which are in A from n until the line BC; but how can squares of finite terms be infinite?[42]

Fontenelle accepts this paradox finally[43] for two main reasons.[44] The first one appeals once more to obscurity[45] and to the special dynamic governing the passage from the finite to the infinite.[46] The second one, more suggestively, relies on the internal coherence of the system and the fruitfulness of the paradox mentioned, in the sense that this paradox, once admitted, "never leads to any false conclusion. To the contrary, it is necessarily bound up with truths already known, and produces many new ones from them. One will be fully convinced of this in what follows." As a result, if the paradox is in fact false, it must nonetheless be "perfectly equivalent to something true" and "very happily" take the place of it. It is therefore proper "in awaiting this true thing" to "accept this paradox as a truth, [as] demonstrated in the preceding art[icle], while nonetheless standing ready, & I say it with the utmost sincerity, to reject it absolutely, once I am made to see that without employing it one can make a System bound up with the Infinite in geometry, or that there is some other idea to be substituted for it which produces the same effect without having the same difficulty, or something equivalent [to this]."[47]

Given this, Fontenelle argues for giving the name "*indeterminable Finites* [to] the finite terms of A that become infinite in A^2 by being raised to squares: for as they are [found] in the passage that A^2 makes from the Finite to the Infinite, they can never be either known or determined as the terms which are at the beginning of A and of A^2 [can be]."[48]

Next he generalizes these results to cases of whole and fractional powers of A.[49]

In short, Fontenelle's study of the series A of the natural numbers led him to regard it as comprising three great sets of elements: determinable finites, indeterminable finites, and indeterminate infinites. To better illustrate the complex situation arising from this division of elements in A, let us take the time to examine a very long excerpt from a letter addressed in April 1729 by Christophe Bernard de Bragelongne (1688–1744), a particularly zealous disciple of Fontenelle, to Daniel Bernoulli (1700–1782) in reply to the letter of 5 October 1728 that Bernoulli had written to Fontenelle after having received a copy of the *Elements*. Bragelongne writes:

> 7°. That being granted, if one takes m to represent all the determinable finites and n to represent all the indeterminable finites, [and] if beyond that (for lack of a large enough number of different characters) one represents the infinites that make up the second and largest part of the series by fractions whose numerators [bear] always the Characteristic ∞, and the denominators, successively, as they are removed from the last term, the determinable finites and then the indeterminable finites, such that these denominators always decrease in approaching the last term ∞, until becoming $= 1$, it is obvious that the series marked A will adquately represent the changes that occur in the Series of natural numbers before arriving at its last term, which is always ∞.

$$\text{A. / } 0,1,2,3,4,\&c \overset{B}{..} m ../.. \overset{C}{n} ../.. \overset{D}{..} \frac{\infty}{n} ../.. \frac{\infty}{m} .. \frac{\infty}{100} .. \frac{\infty}{4} .. \frac{\infty}{2} .. \overset{E}{\frac{\infty}{1}} \text{ /}$$

In fact one can conceive of all the determinable finite numbers as being contained [in the series running] from A to B, the indeterminable finites from B to C, and finally this infinity of infinite terms which goes on always increasing will extend from C until E. It is for these sorts of infinites that Mr de Fontenelle adopts the Characteristic $\propto$, when they have not yet arrived at their last term; I represent them here by different fractions, to be able to distinguish them from each other, without nonetheless abandoning the Characteristic of the Author.[50]

The introduction in Fontenelle's scheme of indeterminable finites and indeterminate infinites thus gives a certain depth, as it were, to the "obscure passage from the finite to the infinite" within which the apparent contradictions of the new calculus in fact resolve themselves. Or, to quote Léon Brunschvicg, it is "through the distinction between the obscure dynamism of the passage to the infinite and the inherent clarity of the static idea of the infinite" that "the apparent contradictions of the calculus of the infinite" are resolved.[51]

As for the "infinitely small magnitudes,"[52] they are only, to use one of Fontenelle's expressions, infinitely great magnitudes "inverted."[53] And, correspondingly, the concepts of indeterminable finite and indeterminate infinite are seen to be altogether naturally introduced in this new context.

Then, after having developed in the very long section VII ("On Infinite Series of Any Magnitude Whatever")[54] a number of considerations centered particularly on the question of the "sums of these series," Fontenelle comes finally to the case of "series whose differences are infinitely small."[55] Study of these series, whose "variations will be managed by infinitely small degrees,"[56] leads Fontenelle to introduce, for the purpose of expressing his concepts, a notation which is nothing other, in fact, than the Leibnizian notation:

> If one calls y any variable magnitude whatever, whose perpetual increase or decrease, regulated by some Law, forms a Series, one will call dy its infinitely small differences. These dy are $\frac{1}{\infty}$, & in fact these are fractions or infinitesimally small parts of y, & dy is $\frac{y}{\infty}$. If the dy are variables, they will also have differences which will be by comparison with them as are the dy by comparison with the y. One calls them ddy, & [so] one has ddy $= \frac{dy}{\infty}$. Similarly the ddy will be able to have their dddy, &c.[57]

Thus, for the first time in Fontenelle's work, the Leibnizian calculus is linked to his inquiry into its foundations and in the process acquires a new clarity and, if one is to believe the author, indeed a new rigor. Moreover, Fontenelle adds at the end of section X ("On the Variations and Changes of Curves"):[58]

But there is always in the knowledge of Curves, even geometric [ones], a great number of things which demand the Theory of the Infinitely large or small, at least to be known in a general way & [one] common to all Curves, & at the same time drawn directly from the bottom of their nature. It is therefore also necessary to have the Art of calculating the Infinites that enter into these Curves, & above all the Infinitely small, because, as we have seen, the infinitely small differences of finite magnitudes that enter into them are the most important [ones] to consider. For this very reason, this Calculus is called *Differential.*[59]

However, it is really only beginning with section XI of the *Elements* ("General Rules for Determining by the Differential Calculus That Which Belongs to the Course of a Curve Plotted against an Axis")[60] that Fontenelle systematically uses the Leibnizian notation to express his own concepts. The decisive point here is that to different orders of differentials there correspond different orders, in Fontenelle's sense, of "absolutely infinitely small" magnitudes. The end of the work is therefore devoted mainly to reorganizing the whole body of results, including those pertaining to motion, so far obtained by his contemporaries with the aid of Leibniz's calculus.[61] The question of the meaning and status of the new mathematical physics was thus now at last posed.

2. Mathematical Physics and the Rationalization of Infinites

From his close reading of the many papers by Varignon devoted to the science of motion, among others, Fontenelle had succeeded in grasping the special character of mathematical physics and of the work of the mathematical physicist. Certain passages drawn from his introductions to the annual volumes of papers of the Royal Academy and from the funeral orations he composed for his colleagues show this more clearly and forcefully than any commentary can.

The secretary of the Royal Academy called particular attention to the "rules," "laws," and "general formulas" found, for example, in Varignon's work, which made it possible both to derive by means of well-determined procedures various scattered results that previously lacked any obvious connection with one another, arranging them now within an ordered theoretical framework, and to open up new lines of investigation insofar as a given "rule," being well demonstrated, now became itself a point of departure. Thus, in the eulogy he composed following the death of his friend Varignon, Fontenelle says:

> In these mathematical researches his genius led him always to make them more general than had [hitherto] been possible. A landscape all of whose parts will have been seen one after the other has not for all of that been seen; it needs to be [viewed] from a high enough place, where all the objects [that] previously [were] dispersed are brought together in a single glance. It is the same with geometrical truths: one can see a great number of them scattered about here and there, without order among them, without connection; but to see them all together and in a glance one is obliged to climb up to a great height, and this demands effort and skill. The general algebraic formulas are the elevated places where one positions oneself to discover a great country all at once. There has perhaps not been any geometer who has either better understood these formulas, or better brought out their value, than M. Varignon.[62]

Or again, a bit later in the same eulogy:

> All of the volumes published by the Academy record his researches. They are almost never [broken up in] pieces, each one detached from the other, but [involve] whole great theories on the laws of motion, on central forces, on the resistance of media to motion. There, by means of his general formulas, nothing that falls within the bounds of the matters which he treats escapes him. Besides new truths, one also sees others already known, but [in a] detached [way], which come from all sides to enter into his theory. All together [they] make up [a] body, and the voids that they left before among themselves are [now] filled.[63]

Very similar ideas are also presented with the same sharpness in the "historical parts" of the annual volumes of the Royal Academy:

> The art of these sorts of formulas consists in taking the thing in its first sources, in what makes its essence, & always subsists, whatever may be the differences which may otherwise survive in it. The question once [having been] raised to its most universal terms, it [remains] only to lower it to particular cases. One finds an algebraic equality which contains nothing determined, only a certain fixed & variable relation; all the rest which depends on the infinite diversity of the different applications [of the calculus] is expressed only in an indeterminate manner, & for which one can substitute [any] such particular & determinate relation as one pleases.

Often Problems which have caused a deal of trouble for great Geometers become very easy small corollaries of these *general formulas*. For example, in the matter of the Clepsydras, if one asks the figure of a Vessel, or, given Galileo's proportion for the speed of outflow, [whether] the surface of the water descends equably in equal times, the formula of Monsieur Varignon gives this figure at once, which it seems that Toricelli was not able to find, & that Monsieur Mariotte was able to find only by a method limited to this particular case.[64]

Fontenelle's attention to "laws," "rules," and "general formulas" allowed him to grasp the peculiar meaning of the work of the mathematical physicist. This consisted to a large degree in perfecting a mathematical instrument in such a way that it brought different domains nearer together, either through the recognition of identical structures or by the establishment of deductive links between structures. The mathematical instrument thus managed to transcend its instrumental status to acquire an explanatory value. Fontenelle was the herald of a mathematical physics that was explicitly aware of both its inductive value and its demonstrative power:

On Central Forces

What has been reported of M. le Marquis de l'Hôpital, & of M. Varignon on central Forces in the Hist. of 1700 seems to have exhausted this matter, & it would be difficult to imagine in relation to [such] Forces some geometrical inquiry of which this Theory does not furnish the principles & the Solution. Those who have pushed it to this point claim to be indebted to the Geometry of Infinitely small [differences], & do not believe that any other Method could attain it. But M. Varignon has not contented himself with showing that this Method is the only one that may lead to such high speculations; he has wished to prove further that it has several paths to get there, & [that] it finds again here by new, & entirely different, routes the same truths that it had already demonstrated. That can only be taken, if you will, as praise for the fruitfulness of the [method of] Infinitely small [differences], although one could claim with reason enough that one always perfects Geometry in presenting the same Truths from as many different sides as possible. Demonstrations of the same subject drawn from entirely different principles are in some way new Instruments of knowledge, & new Organs which one gives to the Mind to grasp an object, & to be sure of it.

Any Curve whatever being conceived as enveloped in all its extent by a string, if one takes one of the extremities of this string, & extends it in a straight line in unwinding it, in such a way that by its other extremity it is always a Tangent of the Curve, it will describe by its first end another Curve, in relation to which the first [curve] is called the *Evolute*. The portion of the string contained between any point of which it is Tangent to the Evolute, & the corresponding point where it terminates at the new Curve, is called the *Radius* of the Evolute, & this name Radius is all the more appropriate as one can effectively consider this portion at each step that it takes as describing an infinitely small arc of a circle, which is a part of the new Curve, [the whole of which is] composed of an infinity of these arcs, all described from different centers, & with different radii. Thus the Radius of the Evolute is always perpendicular to the new Curve, while it is always Tangent to the first.

Every Curve can therefore be conceived as formed by the evolution of another, & it is necessary to find which [curve] is the one whose evolution formed it, which reduces to finding the Radius of the Evolute at any point; for as it is always tangent to the Curve which may be called generative [of it], it is properly [regarded as] only one of its infinitely small parts, or one of its produced Sides, & all these Sides, whose position is consequently determined, are nothing other than the generative Curve itself. This result about the Radii of Evolutes, very useful for the advanced Speculations of Geometry, is [useful] sometimes also for practice, as the application of it that *M. Hugens,* the first Author of this whole idea of evolutions, made to the Pendulum well proved.[65]

The Geometry of the Infinitely small finds always in a general way, in making use of its infinitely small magnitudes, a whole kind of finite magnitudes, [of] whatever variety they may be capable of receiving. Thus all the Tangents of all the possible Curves at all their points are contained in the single relation of the infinitely small [difference] of an Abscissa to that of the corresponding [ordinate]. When the relations that one seeks are not in the infinitely small [differences] of the first type, one has only to press on further, & one finds them. There is therefore a relation of infinitely small [differences] which gives at once the Radii of the Evolutes of all possible curves, & one can see it in the Book of M. le Marquis de l'Hôpital.[66]

Mr. Varignon found the means of passing from knowledge of the radius of the Evolute to that of the central Force, such that when one has the radius of the Evolute of any Curve, one may obtain the value of the central Force of a Body which, being moved by this Curve, will find itself at the same point where this radius terminates, or conversely that by the central Force one may obtain the radius of the Evolute: all this through different relations of infinitely small [differences].

But in unifying these two Theories, he extended them both.

The whole Spirit, & in some manner the Essence of these sorts of Methods, consists in Formulas, or geometrical expressions. In these Formulas, always composed of the relation of several magnitudes, there are some that must remain constant, & invariable, while others may vary as one pleases; they are as the fixed points to which motions are related, & the formulas are all the more general as there are more magnitudes [that are] variable at will, & fewer constant magnitudes. In this regard, the Formulas of M. Varignon for the radii of Evolutes & for central Forces are the most general that may be conceived, since everything in them is indeterminate, they are subject to no constant magnitude that limits them, & they present, so to speak, an infinite space on all sides where everything can be received.[67]

Other, briefer excerpts repeat these same themes while, thanks to their clarity of expression, formulating them in very suggestive ways:

[. . .] the general Rules of M. Varignon [. . .] go infinitely further than all the known phenomena of either astronomy or physics, and one may say that in this matter geometry is presently in the process of resolving more questions than nature will furnish. It does not seem possible to imagine anything about central forces which is not comprehended by the theory of M. Varignon, and so it is a subject that may henceforth be put aside as exhausted.[68]

When one takes the questions of Geometry in more general terms, & that one may embrace within a single Problem a greater view, one always extracts the fruit of it, either in discovering new truths, or in seeing the links and mutual connections of truths already known, or at least in perfecting the Art which considers them, & in giving a greater scope to the Instrument which must grasp them.[69]

This is the use of the exactitude of Geometry. It gives us in all its purity the Truth, which Physics & experiments always alter, & it makes us see how far we, who cannot help but make mistakes, make mistakes with impunity.[70]

Thus, for Fontenelle, mathematical physics draws its power from the fact that it must be governed by a requirement of total mathematicity, for it is only at this price that truly assured knowledge can be acquired and that such knowledge, by virtue of this very requirement, is capable of moving forward.

Fontenelle did not fail, moreover, to assume himself the tasks of the mathematical physicist. In fact, in the eighth section of the second part of his *Elements*,[71] he attempts on the basis of his "general system of the infinite" to give a deeper meaning to the concepts of "speed in each instant" and "accelerative force in each instant." Here he tries in particular to resolve the difficulties relating to the beginning and continuity of motion:

> 1542. In the 1st instant of the fall of bodies, the speed has only been [an] infinitely small [quantity] of the 1st order, since a space dde has been traversed in a time dt. Next, 3dde being traversed in an equal 2nd dt, [then] 5dde in a third dt etc., the speed remains infinitely small, until [the point] that the coefficient of dde is α, for then α dde = de being traversed in a dt, the speed is finite. Now the coefficient which expresses the number of dde traversed in a dt being always a term of the series of odd numbers, it can be infinite only when this series has had an infinite number of terms, which is easy to see, or after it has had an infinite equal number of dt, that is to say, after a finite time. Therefore, it is only after a finite time that an infinitely small space de of the 1st order comes to be traversed in an instant dt of the same order.
>
> 1543. Therefore, the accelerative force requires a Finite time to impress upon a body a finite speed, or, what is the same, it needs to be applied during a finite time.[72]

Fontenelle attempted also, among other things, to deduce from the principles of his system the conditions under which a varied motion can be treated (as Varignon, for example, assumed it could be) as a succession of uniform motions, each taking place in an instant of time:

> 1548. As in each instant an equal, infinitely small speed is always added to the sum of speeds already acquired, the augmentation is of the same order as the sum so long as the sum is only an infinitely small [quantity]. But this sum is an infinitely small [quantity] as long as the total time of the fall of the body or of the total distance traversed is still not finite (1543): but from the instant at which this happens [i.e., when the distance traversed becomes finite], the speed of each following instant being finite, the augmentation of speed is no longer to be counted in relation to it, and one can take the speed of each instant as uniform, and express it as such by the relation of the distance de of this instant to dt. Thus it is only by supposing a finite time already passed, and a finite distance traversed, that one can take the speed of each instant as uniform.[73]

Through this analysis, or more exactly through the meaning of this analysis, Fontenelle clearly indicated, on the one hand, that the concepts of the science of motion must be subject to the same criteria of rigor as the mathematical concepts on which they depend and, on the other hand, that this analysis of the concepts of the science of motion, so far as it is governed by a requirement of total mathematicity, can lead to a better understanding of the content of these concepts and of the conditions under which they are to be used.

The notion of the infinite, even if it had not yet been mastered in a strict sense, nonetheless seemed to Fontenelle, in the context of his distinction between geometrical and metaphysical infinity, as capable of being mastered and thus of clarifying, indeed of actually grounding, the meaning of mathematization and, correspondingly, that of mathematical physics. Fontenelle's failure was therefore a matter of failing to succeed in rationalizing the infinite—of failing, in the early eighteenth century, in the quest for the meaning of the new mathematical style of physics associated with a longing for true knowledge.

In attempting to rationalize the infinite, the work of Fontenelle constitutes, as the passages cited above testify, a particularly significant reflection upon the question of the original meaning of mathematization and upon its development. In fact, Fontenelle seems, on the one hand, convinced of the necessity of constructing a meaningful rationality of the infinite and, on the other, fascinated by the possibility of doing this "in the Varignonian fashion," that is, without inquiring into the question of meaning, into the status of formulas and rules capable of summarizing or reducing a phenomenal multiplicity in a few symbols.

Fontenelle's thought—the object of the *Élemens de la géométrie de l'infini* being precisely to propose a way of getting around this tension—was therefore both pregnant still with the original sense of mathematization and at the same time already filled with a forgetfulness of this sense in its embrace of the pure investigation of formulas and rules associated with the reduction of mathematics to simple language, so that the coherence and the value of information brought by mathematics to the study of phenomena came in the end to be ignored.

The failure of Fontenelle's project, combined with the successful founding and development of a new science of motion by his contemporaries (among them Jean Bernoulli, Leonhard Euler, Alexis-

Claude Clairaut, and Jean d'Alembert, to name only a few), thus implied an extreme valorization of the technical dimension of mathematical manipulation insofar as it allowed a great many practical applications to be obtained from formulas; and this coincided definitively with a forgetting of the original sense of mathematization "as completing the knowledge of the world," in Husserl's expression.[74]

Regarded in this perspective, the *Mécanique analitique* published in 1788 by Joseph-Louis Lagrange was exemplary. Lagrange was moreover fully aware of the fact that his work constituted both a model and an end point. Several lines taken from the foreword to the first edition are exceedingly revealing: "I have proposed to reduce the theory of this Science [mechanics], and the art of resolving the problems that relate to it, to general formulas whose simple development gives all the necessary formulas for the solution of each problem [. . .]. Those who love Analysis will take pleasure in seeing mechanics become a new branch of it, and will be grateful to me for thus having extended its domain."[75]

Ernst Mach made a point of stressing the importance of Lagrange's achievement in his celebrated critical and historical account of the development of mechanics: "Analytical mechanics [. . .] was brought to its highest degree of perfection by Lagrange [in *Mécanique analytique*]. Lagrange's aim is [. . .] to dispose *once and for all* of the reasoning necessary to resolve mechanical problems, by embodying as much as possible of it in a single formula."[76]

Lagrange's work is divided into two parts, the first devoted to "Statics," or the "science of the equilibrium of forces" (it being understood that Lagrange "understands, in general, by *force* or *power* the cause, whatever it may be, which impresses or tends to impress motion upon the body to which it is supposed to be applied");[77] and the second devoted to "Dynamics," or the "science of accelerative or retardative forces, and of the varied motions which they must produce."[78]

After having treated "different principles of Statics," mainly from a historical point of view, in the first section of the first part, Lagrange proceeds in the second section, evocatively titled "General Formula of Statics for the Equilibrium of Any System of Forces, with the Manner of Employing This Formula,"[79] to consider the particular object of his investigation. Lagrange takes as his point of departure in statics the "principle of potential speeds," which he states as follows: "The general law of equilibrium in machines is that the forces or powers are among themselves inversely as the speeds of the points to which they are applied, reckoned according to the direction of these powers."[80] Therefore, "to reduce this principle to a formula," let the forces or powers "P, Q, R, etc. [be] directed along given lines" in equilibrium. Next, from the points where these powers are applied, one draws "straight lines equal to p, q, r, etc. and oriented in the directions of these powers." One then designates, "in general, by dp, dq, dr, etc. the variations, or differences of these lines, insofar as they may result from any infinitely small change in the position of the different bodies or points of the system." The displacements involved here are infinitely small, and as a consequence, according to Lagrange, "it is clear that these differences will express the distances traversed in a single instant by the powers P, Q, R, etc. traveling in their own directions, assuming that these powers tend to augment the respective lines p, q, r. The differences dp, dq, dr, etc. will thus be proportional to the potential speeds of the powers P, Q, R, etc. and [therefore] can, for greater simplicity, be taken as [equivalent to] these speeds."[81]

In a first case, the only one that we shall examine here, albeit cursorily, Lagrange considers only two forces P and Q in equilibrium and then applies the law of equilibrium. In this situation, "it will be necessary that the quantities P and Q be among themselves in [the] inverse ratio of the differentials dp, dq [. . .], whence it follows that the values of the differences dp and dq must be of opposite signs: the values of the forces P and Q therefore being assumed both [to be] positive, one obtains for equilibrium

$$\frac{P}{Q} = \frac{dq}{dp}, \text{ or } Pdp + Qdp = 0."$$

This result is "the general formula of equilibrium for two powers."[82] Lagrange then generalizes it to the case "of any number of powers P, Q, R, etc." and arrives at the result so long and much awaited, the object of all the care and point of the exercise, namely, "the general formula of Statics": "One has therefore, in general, for the equilibrium of any number of powers P, Q, R, etc., directed along the lines p, q, r, etc. and applied to any system of bodies or points arranged among themselves in any manner, an equation of the form:

$$Pdp + Qdq + Rdr + \ldots = 0$$

This is the general formula of Statics for the equilibrium of any system of powers."[83]

This general rule in hand, nothing more was needed but to specify the rules for putting it to proper use, that is, for resolving the problems of statics by means of "analytical operations" alone: "The object of this work being to reduce Mechanics to purely analytical operations, the formula that we have just found is well suited to achieve [this result]. It is a question only of expressing analytically, and in the most general manner, the values of the lines p, q, r, etc. taken in the directions of the forces P, Q, R, etc., and one will have, by simple differentiation, the values of the potential speeds dp, dq, dr, etc."[84]

The first part of the *Mécanique analytique* having been devoted to statics, the second, as we have said, is devoted to dynamics. This second part, mirroring the first, opens with a section based essentially on a historical approach to "different principles of dynamics." The section that follows concerns a "General Formula of Dynamics for the Motion of a System of Bodies Impelled by Any Forces Whatever."

Lagrange defines "the forces immediately spent in moving the body m (mass) during the time dt, parallel to the axes of the coordinates xy, y, z," in a very general form that was to become standard, namely,

$$m\,\frac{d^2x}{dt^2}, \quad m\,\frac{d^2y}{dt^2}, \quad m\,\frac{d^2z}{dt^2}$$

He next considers each "body m of the system as driven by similar forces" and concludes from this that the sum of the moments of these forces, represented by the formula

$$\int \left(\frac{d^2x}{dt^2}\,\delta x + \frac{d^2y}{dt^2}\,\delta y + \frac{d^2z}{dt^2}\,\delta z \right) m$$

(where "the variations which express the potential speeds" are denoted by the "character δ," and "assuming that the sign of integration ∫ is extended to all the bodies of the system"), must constantly balance, or equilibrate, the sum of the moments of the "given accelerative forces (P, Q, R, etc.) which attract each body m of the system, toward the centers to which these forces are supposed to tend."[85] One obtains then

$$\int\left(\frac{d^2x}{dt^2}\,\delta x\ +\ \frac{d^2y}{dt^2}\delta y\ +\ \frac{d^2z}{dt^2}\,\delta z\right)m\ +\ \int$$

$$(P\delta p\ +\ Q\delta q\ +\ R\delta r\ +\ ...)m\ =\ 0$$

As in the case of statics, this result "is the general formula of Dynamics for the motion of any system of bodies."[86] It is thus easily compared with that derived in the first part of the book: "It can be seen that this formula differs from the general formula of Statics [. . .] only by the terms due to the forces

$$m\frac{d^2x}{dt^2},\ m\frac{d^2y}{dt^2},\ m\frac{d^2z}{dt^2}$$

which produce the acceleration of the body m along the extensions of the three coordinates x, y, z [. . .]."[87] The solution of the problems of dynamics therefore reduces, as in the case of statics, to "analytical operations" alone.

In this sense, Lagrange's *Mécanique analytique* amounted to an extraordinary extension of the "Varignonian style" to the whole of mechanics. The Lagrangian completion of what Varignon inaugurated was accompanied, as Fontenelle uneasily suspected, by a forgetting of the original meaning of the process of mathematization and of the generation of mathematical physics.

Lagrange's results, just like those of Varignon, though to a lesser degree, by holding forth the possibility (with a stunning and, if one may so so, "mechanical" ease) of resolving a multiplicity of problems on the basis of one or two general formulas, using purely analytical procedures, made these formulas the object of great fascination, to the point that their derivation came to be regarded as the purpose of scientific endeavor. Meaning came to be concealed by method. Thus the original meaning of mathematical physics as a discipline, which, it was understood, would affirm itself in completing the knowledge of the world,[88] vanished in the stubborn search for formulas that could serve as points of departure for algorithmic calculation. Thus, too, the modern notion of the formula came to find its harbingers and philosophical excuses.

Ernst Mach, to name only one example, was to insist on the

utilitarian and instrumental character of science, utility being construed in a broad sense since, for Mach, it was part and parcel of
a larger economy of thought: "It is the object of science to replace,
or *save,* experiences, by the reproduction and anticipation of facts
in thought. Memory is handier than experience, and often answers
the same purpose. This economical office of science, which fills its
whole life, is apparent at first glance; and with its full recognition
all mysticism in science disappears."[89]

On this view, Lagrange's work appeared altogether exemplary.
A further indication of its influence upon physicists such as Mach
may be had by considering at greater length a passage from his
Die Mechanik (1883) cited in part a few pages earlier: "Analytical
mechanics [. . .] was brought to its highest degree of perfection by
Lagrange [in *Mécanique analytique*]. Lagrange's aim is [. . .] to dispose *once and for all* of the reasoning necessary to resolve mechanical problems, by embodying as much as possible of it in a single
formula. This he did. Every case that presents itself can now be
dealt with by a very simple, highly symmetrical and perspicuous
schema; and whatever reasoning is left is performed by purely
mechanical methods. The mechanics of Lagrange is a stupendous
contribution to the economy of thought."[90]

Science, from the point of view of the new mathematical physics
codified by Lagrange's great work of 1788, was therefore impelled
on its way along a road that led away from meaning. The reasons
for this forgetfulness were queried, and queried again, because
only through this ceaseless process of questioning and reexamin-
ation might the torn bonds of meaning one day be reknit.

Thus, just as Fontenelle had questioned the work of Varignon,
so too the mathematician Hoené Wronski (1776–1853), the first
advocate of Kant's philosophy in France, questioned that of Lagrange. His was a new style of questioning, a transcendental style
which, as sketched in a collection of Wronski's essays published
in 1814, *La philosophie de l'infini,* inaugurated in some sense the
tradition of interrogating modernity.[91]

In the essay "Philosophie du calcul infinitésimal," Wronski re-
visits the question of the foundations of the calculus from a perspective that is opposed both to that of Lagrange and to that of
Lazare Carnot[92] but which hooks up again, with regard to the
status of infinity, with that of Fontenelle. Wronski's philosophical
ambitions were quite clear: he wished to give the differential and

integral calculus a strictly speculative foundation, relying for the most part upon a Kantian transcendental analysis. The opening lines of this essay are remarkable, not least for their lack of ambiguity:

Before everything, it must be recognized that the idea of the INFINITE is an intellectual product utterly different from that constituted by the conception of a FINITE quantity. These are two functions of our knowledge [that are] utterly heterogeneous. The one, the conception of a finite quantity, is a product of the UNDERSTANDING, which serves to tie together intellectually the intuitions that we have of objects, or, if you like, of these objects themselves; that is to say, to speak a more philosophical language, the conception of a finite quantity is a product of the Understanding, which, under the conditions of time that are specific to it, introduces an intellectual unity or meaning in being [as] opposed to learning. The other of the two functions [in] question, the idea of the infinite, is a product of REASON, which, in itself, stands outside the conditions of time, and consequently is inapplicable or transcendent [with respect to] the constitutive use that we make of learning for the knowledge of being, that is to say, inapplicable [to] that particular use of learning that constitutes the laws of our immanent knowledge, or of [that part of] our knowledge which can be established by experience, whose first condition is time. But, employed at least in a regulative manner, in [being] submitted by the influence of JUDGMENT to the conditions of time, which are foreign to it, the idea of the infinite, thus transformed into the idea of the INDEFINITE, serves to tie together the very conceptions that we have of quantity; that is to say, speaking again in a more philosophical language, the idea of the infinite, through which [breathes] the ABSOLUTE, finding itself transformed by virtue of the mediation of Judgment into the idea of the indefinite, through the application of the conditions of time, operates at least regulatively in the immanent sphere of our knowledge, [by] introducing the ultimate unity or the ultimate significance, not in the object of knowledge, [i.e.,] in being, but rather in the very functions of knowledge relating to the knowledge of quantity. Thus, and this is what needs to be noted here, the conception of a finite quantity bears always upon the object of knowledge, upon being, which is opposed to knowledge and which constitutes the object of knowledge; whereas the idea of the infinite, which, by itself, stands outside the conditions of time and which, consequently, finds no immediate point of application to the object of knowledge, or being, which is opposed to it, cannot in [being] submit[ted] to the conditions of time through Judgment, [be] use[d] in the sphere of our immanent knowledge, that is to say, in [being] transformed into the idea of the indefinite, can only, [being] discontinuous, bear upon the very functions of knowledge, where it introduces the highest intellectual unity and the highest significance in the very production of the knowledge of quantity. In a word, the conception of a finite quantity serves as [a] constitutive law of the possible relations in being [as] opposed to knowledge; and the idea of the infinite, transformed into the idea of the indefinite by the application of the conditions of time, serves only as [a] regulative law or rule [regarding] the very function of knowledge concerned with the generation of knowledge about quantity.[93]

This transcendental distinction, resting on the two heterogeneous functions of knowledge, understanding and reason, served as the basis for Wronski's elaboration of a philosophy of the infinitesimal calculus in the sense that "finite quantities bear upon the objects of our knowledge, and infinitesimal quantities on the very generation of this knowledge; such that each of these two classes of knowledge must have laws proper [to them], and it is in the distinction between these laws that the major thesis of the metaphysics of infinitesimal quantities is to be found."[94]

Two types of laws were thus distinguished by Wronski: those relating to finite quantities are called "objective" laws, and those relating to infinitesimal quantities are called "subjective" laws. The force of this was to say that "the inexactitude that is felt to be attached to the infinitesimal Calculus"[95] resides in the confusion of the two types of laws:

> To better distinguish these laws, we name the laws of finite quantities *objective* laws, because they bear upon the objects of our knowledge; and the laws of infinitesimal quantities *subjective* laws, because they bear only upon the generation of our knowledge relating to quantity.
>
> Now, the first scientific result that we obtain from this transcendental distinction is the NEGATIVE PRECEPT of not confusing, in Algorithmy, the objective laws of finite quantities with the purely subjective laws of infinitesimal quantities. It is this confusion that is the source of the inexactitude that is felt to be attached to the infinitesimal Calculus; in fact, [this amounts to] confusing the subjective laws of infinitesimal quantities, which are only rules of our speculation about the generation of knowledge of quantity, with the objective laws of finite quantities, which are rules of the very reality of quantity, and it is thus that one is naturally led to confuse them, [for] one believes discover[ed] in the procedures of the infinitesimal calculus a kind of

logical contradiction, or even of absurdity, arising, as one now sees, from the transcendental antinomy that is found between the products of Reason and those of the Understanding. This is also [why] geometers, especially those of the present day, consider the infinitesimal Calculus, which nonetheless they concede always gives true results, to be only an indirect or artificial procedure, or at least to be founded on different principles than the simple principles from which the scientists who discovered it deduced it.[96]

Among the scientists who attempted to give an account of the infinitesimal calculus while interpreting it, since they were ignorant of the transcendental distinction, as a simple indirect or artificial procedure, a mere technique of calculation (an interpretation that Fontenelle had inveighed against earlier), Wronski places Lagrange and Carnot first and foremost.

Without entering into the mathematical detail of Lagrange's work, which owes so much to the introduction of algebraic methods, merely reading the title of his 1797 treatise on these questions is, from the perspective of Wronski's analysis, illuminating: *Theory of Analytic Functions, Containing the Principles of the Differential Calculus, Freed from All Consideration of Infinitely Small or Vanishing Quantities, Limits and Fluxions, and Reduced to the Algebraic Analysis of Finite Quantities.*[97]

As for Lazare Carnot, his approach, which on many points recalls that of George Berkeley—an approach relying on the idea that the errors due to infinitesimals had to be eliminated, and therefore to be made to disappear at the end of a calculation—could hardly satisfy Wronski's infinitist and transcendental perspective. A few lines from Carnot's *Reflexions on the Metaphysical Principles of the Infinitesimal Calculus,* also published in 1797, are enough to suggest what might have prompted Wronski's critique:

> Moreover, we may, instead of any quantities whatsoever contained in them, substitute others differing in an infinitely small degree; we may neglect infinitely small quantities relatively to finite quantities, and in more general terms "accessary" *quantities,* as compared with principal quantities; without these equations ever losing their primitive character of imperfect equations, and which become exact eventually by a compensation of errors.
>
> But it is worth while to remark, that these continued errors, instead of removing us farther and farther (as it apparently must) from our object, (which is to reduce these imperfect equations to perfect exactness,) tend on the contrary to lead us to it by the shortest and simplest method, since by thus getting rid of these

inconvenient "accessary" quantities in succession only, taking care not to affect the truth of the proposed equations, we at least disengage them from every consideration of infinity by the entire elimination of all arbitrary quantities, and observing that these quantities alone, the ratio of which we wished to know, be suffered to remain. This being laid down, the whole theory of infinite quantities may be comprised in the following theorem.

Theorem.

34. For an equation to be necessarily and rigorously exact, it is enough to know,

1. That it has been deduced from true or "imperfect" equations at the least, by transformations which have not deprived them of the character of imperfect "equations."

2. That it no longer retain any Infinitesimal quantity, that is to say, any other quantity than those of which it was proposed to find the ratio.[98]

In each case, with Lagrange as with Carnot, if one accepts Wronski's reading, attention to the infinite has been abandoned, as a result of confusing objective and subjective laws, in favor of finite quantities.

Owing to the distinction that, from a transcendental standpoint, exists between finite and infinitesimal quantities, the celebrated "Demand" formulated a century earlier by the Marquis de l'Hospital in his celebrated *Analyse* of 1696 became, contrary to all expectation, the "positive precept" of Wronski's interpretation of the infinitesimal calculus and no longer implied any contradiction:[99]

Having thus avoided the confusion between the objective laws of finite quantities and the purely subjective laws of infinitesimal quantities, as prescribe[d] [by] the negative precept, which is the first result of the transcendental distinction between these laws, it is necessary, to complete the Metaphysics of the infinitesimal Calculus, to deduce the principle of the subjective laws that are the object of this calculus; and it is evidently there [that] the POSITIVE PRECEPT [is found], resulting from the transcendental distinction in question.

Now, this principle of the subjective laws [which] constitutes the object of the infinitesimal Calculus, is nothing other than the grand principle of the infinitesimal Calculus itself, namely, [. . .] TWO QUANTITIES THAT DIFFER AMONG THEMSELVES ONLY BY AN INDEFINITELY SMALLER QUANTITY ARE STRICTLY EQUAL.[100]

Rigor was thus reintroduced and the "Demand" made principle. This principle, clearly understood now as "a subjective rule for generating knowledge about quantity,"[101] was seen to possess an "apodictic certainty." It could also be the object of a true "meta-physical deduction":

> Here, moreover, is the rigorous metaphysical deduction of this grand principle.
>
> Since, as we noticed earlier, the laws of infinitesimal quantities are purely subjective, that is to say, they are only rules for generating knowledge about quantity, and not objective laws of the relation between quantities itself, it is immediately true, and this not only intuitively, through a synthetic a priori judgment, as with other principles of Mathematics, but discursively as well, through the logical principle of contradiction alone, it is true, we say, that two quantities A and B which differ among themselves only by an INDEFINITELY smaller quantity C are strictly equal. For the idea of the infinitesimal quantity C being only a rule for generating knowledge about quantities on the order of those which find themselves in relation here, namely, on the order of the quantities A and B, and not actually knowledge acquired or generated about some quantity, because, as we have seen, the indefinitely smaller quantity C does not, either itself or any of its parts, have any objective reality in the sphere of magnitude where the quantities A and B are found, it is clear that the relation of the quantities A and B in question, considered in its objective reality, is not at all changed by the purely subjective influence of the infinitesimal quantity C. Therefore, etc., etc.[102]

Consequently—and this is the whole point of Wronski's analysis—the foundations of the infinitesimal calculus cannot, as Fontenelle tried to do, somehow avoid the "idea of the infinite" by taking refuge in formal procedures and artificial techniques of calculation. Quite to the contrary, this idea, once "subjected to the conditions of time," becomes a "regulative idea" capable of being given a "constitutive usage":

> In concluding this outline of the Metaphysics of the infinitesimal Calculus, we feel justified in expressing the hope that those geometers who will inquire into this metaphysics in greater depth will recognize, no doubt with difficulty, the futility of rejecting infinity. They will understand that the idea of INFINITY, taken in itself, is, in truth, transcendent or immediately inapplicable in the immanent sphere of our knowledge; but that this idea, finding itself subjected to the conditions of time, which is possible through a mediating intellectual faculty, and finding itself thus transformed into the idea of the INDEFINITE, becomes one of the most exact and most powerful instruments of science.

And in fact, this regulative idea gives us, beyond all expectation so to speak, the rules for generating knowledge itself about quantity; which incontestably is the most sublime use of our faculties of knowledge. The geometers [will] also understand the incomparable importance of the infinitesimal Calculus, and above all of the differential Calculus, by comparison with all the other procedures of Mathematics. We think that this repugnance, so strongly established for the idea of infinity, will be overcome all the more easily as, in this way, geometers extricate themselves from the manifest contradiction in which, to their shame, they find themselves sunk, having rejected, on the one hand, the idea of infinity, and having cultivated, on the other, transcendent, irrational quantities of series and a thousand other functions which, without the idea of infinity, signify nothing, absolutely nothing. We have already indicated elsewhere *(Réfutation de la Théorie des fonctions analytiques, à la fin du premier Mémoire)* the source of this logical incoherence among geometers, [the source] of this desire to avoid the idea of infinity, this most sublime instrument of their noblest occupation.[103]

And further:

The infinite not only is an exact instrument of mathematical investigation, but moreover is the most important element of mathematical truths themselves. We say more: IT IS ONLY THROUGH THE INFINITE THAT THE SCIENCE OF MATHEMATICS IS POSSIBLE.[104]

In the period leading from Fontenelle to Wronski, the quest for reasons had been forgotten, in part because of the silence imposed on infinity—forgetfulness and silence that, because they were incessantly questioned, yet managed to impart meaning to the scientific work involved in the development of mathematical physics, above and beyond the practical strategies suggested by positivism.

The inaugural Galilean project, through the very practice of mathematical physics, gradually therefore turned into something different; and this different project crystallized and took on definitive form with the statement of the Lagrangian position in the *Mécanique analytique*. Lagrange's mechanics, once having been adapted and reworked in the nineteenth century by William Rowan Hamilton (1805–1865) and Carl Gustav Jacobi (1804–1851), thus opened up a new era in mathematical physics in the sense that, fortified by Hamilton and Jacobi's modifications, it was to nourish and give order to the whole set of developments in the field that unfolded well into the next century.

If post-Lagrangian mathematical physics no longer adhered to the aims of the Galilean project, the questioning of its foundations, as this continued, for example, with the founding of quantum mechanics between 1925 and 1930 by Niels Bohr, Louis de Broglie, Erwin Schrödinger, Werner Heisenberg, and others, nonetheless raised once more, abruptly and forcefully, the question of the meaning of mathematical physics together with that of the status of the objects of the world.

Galileo's project was therefore no longer that of mathematical physics, not at least as this was practiced on a daily basis. Just the same it continued, and continues today, to furnish the means by which mathematical physics can always be questioned, revised, and renewed; the means by which mathematical physics will always hold out, in its deepest and most vital structure, the possibility of completing our knowledge of the world; and the reason why mathematical physics will always be, in a word, reason.

⎯⎯⎯⎯⎯⎯⎯⎯⎯⎯⎯ ∞ ⎯⎯⎯⎯⎯⎯⎯⎯⎯⎯⎯

Introduction

1. Galileo Galilei, *Il Saggiatore* (Rome, 1623), in *The Controversy on the Comets of 1618,* trans. Stillman Drake and C. D. O'Malley (Philadelphia: University of Pennsylvania Press, 1960), 183–184.

2. Ibid., 184.

3. Quoted in Stillman Drake, *Galileo at Work: His Scientific Biography* (New York: Dover Publications, 1995), 412.

4. René Descartes, *Principia Philosophiae* (Paris, 1644), in *Oeuvres,* ed. Charles Adam and Paul Tannery (Paris: Cerf, 1896–1913) (hereafter abbreviated as AT), 9 (pt. 2). The translation cited here is from *Principles of Philosophy,* in *The Philosophical Writings of Descartes,* trans. John Cottingham, Robert Stoothof, and Dugald Murdoch (Cambridge: Cambridge University Press, 1984–1991), 1:224.

5. Joseph-Louis Lagrange, foreword to *Mécanique analitique* (Paris, 1788); see the complete modern edition edited by Gaston Darboux (Paris: Éditions A. Blanchard, 1965).

6. Gaston Bachelard, *L'activité rationaliste de la physique contemporaine* (Paris: Presses Universitaires de France, 1951), 41.

7. Edme Mariotte, *Essai de logique* (Paris, 1678); see the new edition edited by Guy Picolet in collaboration with Alain Gabbey (Paris: Librairie Fayard, 1992), 168.

8. Giordano Bruno, *De l'infinito, universo e mondi* (London, 1584), cited in Alexandre Koyré, *From the Closed World to the Infinite Universe* (Baltimore: Johns Hopkins University Press, 1957), 44.

9. Thus the question posed by Henry More, for example, referring as well to the Cartesian opposition between the infinite and the indefinite in his *Enchiridion Metaphysicum* (London, 1671), pt. 1, chap. 10, par. 14.

10. Blaise Pascal, *De l'esprit géométrique* (Paris, 1657/58), in *Great Shorter Works of Pascal,* trans. Émile Cailliet and John C. Blankenagel (Philadelphia: Westminster Press, 1948), 195–196.

11. Blaise Pascal, *Pensées,* trans. A. J. Krailsheimer (Harmondsworth: Penguin, 1966), 91; the numbering of the fragments given in the text refers to the Lafuma edition contained in the complete works published by Seuil in 1963.

12. Pascal, *Great Shorter Works,* 201; see also Descartes, AT, 8 (bk. 1, pt. 2, §34).

13. Pascal, *Pensées,* 91–92.

14. Ibid., 149. See also Nicolas Malebranche, *De la Recherche de la Vérité* (Paris, 1675), bk. 3, pt. 2, chap. 2; Ralph Cudworth, *The True Intellectual System of the Universe* (London, 1678), 640ff.; and Girard Desargues, "Brouillon project" (1639), in *L'oeuvre mathématique de G. Desargues,* ed. René Taton (Paris: Presses Universitaires de France, 1951; reprint, Paris: Vrin, 1981), 99, 179.

15. Galileo Galilei, *Opere,* ed. Antonio Favoro, Isidoro Del Lungo, and Valentino Cerruti, 21 vols. (Florence: Barbera, 1890–1909), vol. 8; see Galileo Galilei, *Two New Sciences,* trans. Stillman Drake (Madison: University of Wisconsin Press, 1974), 34.

16. Descartes, *Principles of Philosophy,* in *Philosophical Writings,* 1:201–202.

17. Leibniz, among others, underscored this characteristic of the new calculus. In fact, for Leibniz, in permitting the "transposition of geometry into nature" (*Leibnizens mathematische Schriften,* ed. C. I. Gerhardt [Hildesheim: Georg Olms, 1961] [hereafter abbreviated as GM], 6:151), this calculus did not imply that the mathematical entities corresponding to the concepts of *"conatus,"* attraction, and so on were really found in nature, but that, quite to the contrary, they were only devices for making good estimates by means of abstractions; in this connection see in particular the passage in Leibniz's "Specimen Dynamicum" (1695), GM, 6:238 (or in the new version edited by Hans Gunter Dosch, Glenn W. Most, and Enno Rudolph [Hamburg: F. Meiner Verlag, 1982], 12).

18. *Histoire de l'Académie Royale des Sciences avec les Mémoires de Mathématique et de Physique pour la même année: Tirés des Registres de cette Académie,* 92 vols. (Paris: 1702–1797). Note that each volume corresponds to a single year of the period 1699–1790, being divided into a "partie Histoire" (hereafter *AH*) and a "partie Mémoires" (hereafter *AM*). The passage cited is from *AH,* 1726 (published 1728): 21, 96.

Chapter One

1. Galileo Galilei, *Discorsi e dimostrazione matematiche intorno a due nuove scienze* (Leyden, 1638), in *Opere,* 8:43–313. In what follows, quotations are taken from *Two New Sciences,* with the corresponding passage in Galileo's complete works indicated in parentheses.

2. Galileo, *Two New Sciences,* 109 (*Opere,* 8:152).

3. On these topics see Clifford Truesdell, *Essays in the History of Mechanics* (Berlin: Springer-Verlag, 1968), as well as Truesdell's *The Rational Mechanics of Flexible or Elastic Bodies, 1638–1788: Introduction to "Leonhardi Euleri Opera Omnia," Vols. X and XI, Seriei Secundae* (Zurich: Orell Fussli, 1960); and Stephen P. Timoshenko, *History of Strength of Materials* (New York: Dover, 1983).

4. See Maurice Clavelin, *La philosophie naturelle de Galilée* (Paris: Colin, 1968), esp. chap. 8; or, in English, *The Natural Philosophy of Galileo,* trans. A. J. Pomerans (Cambridge, Mass.: MIT Press, 1974).

5. Galileo, *Two New Sciences,* 75 (*Opere,* 8:116).

6. Ibid., 77 (*Opere,* 8:118).

7. Ibid., 153–154 (*Opere*, 8:197–198). The Latin terms in brackets have been added to Drake's translation.

8. Clavelin, *La philosophie naturelle de Galilée*, 286 n. 4.

9. See Galileo's letter of 16 October 1604 to Paolo Sarpi, in *Opere*, 10:116; see also Alexandre Koyré, *Études galiléennes* (Paris: Hermann, 1966), 86ff. An English edition of the Koyré book is *Galileo Studies*, trans. John Mepham (Atlantic Highlands, N.J.: Humanities Press, 1978).

10. Galileo, *Two New Sciences*, 165 (*Opere*, 8:208).

11. Without entering into a historical discussion that extends beyond the scope of this book, it may be recalled that this result was already known in the fourteenth century. But it goes without saying that its epistemological import was no longer the same, since for Galileo it was a question of naturally accelerated motion. On this point see Marshall Clagett, *The Science of Mechanics in the Middle Ages* (Madison: University of Wisconsin Press, 1959); and Edith D. Scylla, *The Oxford Calculators and the Mathematics of Motion, 1320–1350: Physics and Measurements by Latitudes* (New York: Garland, 1991).

12. Galileo, *Two New Sciences*, 166 (*Opere*, 8:209).

13. See in particular the first section of chapter 3 below, "Satisfying Reason."

14. Galileo concerned himself with these questions in the 1600s. On this point see Stillman Drake's *Galileo* (New York: Oxford University Press, 1980) and *Galileo: Pioneer Scientist* (Toronto: University of Toronto Press, 1990).

15. Galileo, *Two New Sciences*, 175 (*Opere*, 8:215).

16. Galileo, *Two New Sciences*, 162 (*Opere*, 8:205); for a detailed discussion of the sense and scope of this "principle," see Clavelin, *La philosophie naturelle de Galilée*, 366ff.

17. A discussion of these manuscripts may be found in Joella G. Yoder, *Unrolling Time: Christiaan Huygens and the Mathematization of Nature* (Cambridge: Cambridge University Press, 1988).

18. See Christiaan Huygens, *Oeuvres complètes de Christiaan Huygens* (hereafter *O. c. Chr. Huygens*) (The Hague: Société Hollandaise des Sciences, 1888–1950), 17:125–137; and the Huygens manuscripts held by the University of Leyden Library (hereafter Hug.), 10, fols. 80v–86v.

19. From the French translation by Émile Jouguet in his *Lectures de mécanique* (Paris: Gauthier-Villars, 1924), 61–62. It is instructive to compare Jouguet's analysis of Torricelli's result with the Galilean experiment involving two moving bodies that are joined together, the purpose of which was to show that the different speeds with which we see the bodies fall toward the ground (or, more precisely, toward the center of the earth) depend, not on weight in the sense assumed by Aristotelian tradition, but on specific weight (determined by the medium through which the body passes): see Galileo, *Opere*, 8:107–109.

20. See Evangelista Torricelli, *De motu gravium naturaliter descendentium et projectorum* (Florence, 1644), in *Opere di Evangelista Torricelli*, ed. Gino Loria and Giuseppe Vassura (Faenza: G. Montanavi, 1919), 2:104ff.

21. See *O. c. Chr. Huygens*, 16:131–132; Hug. 10, fol. 84v.

22. It will be recalled that the cycloid is a curve described by a fixed point on a circle that rolls along it without falling off onto a line. It had recently been carefully studied, in particular by Roberval.

23. See *O. c. Chr. Huygens,* 16:392–413; Hug. 26, fols. 72r–75r; Hug. 10, fols. 94r–96r.

24. *O. c. Chr. Huygens,* 3:13.

25. Hug. 26, fols. 72r, 73r, 74r; *O. c. Chr. Huygens,* 16:392–403.

26. "Hinc data fuit occasio inventi de Cycloïde. Quaeritur quam rationem habeat tempus minimae oscillationis penduli ad tempus casus perpendicularis ex penduli altitudine" (Hug. 26, fol. 72r; *O. c. Chr. Huygens,* 16:392).

27. In modern terms: $x = \frac{1}{2} at_f^2$ and $v_f = at_f$. Moreover, since $X = vt_f$ it follows that $X = at_f t_f = at_f^2$, and so $X = 2x$.

28. "Tempus per particulam E, ex K cadentis, est ad tempus per particulam B cum celeritate ex AZ in ratione composita ex longitudine E ad B, hoc est ex ratione TE seu GB ad EB [. . .]" (Hug. 26, fol. 72r; *O. c. Chr. Huygens,* 16:393).

29. The infinitely small part (E) is supposed to be traversed with the constant speed reached in E, whereas the infinitely small part (B) is supposed to be traversed with the constant speed v_f obtained by the transformation of the movement in free fall from A to Z in a uniform motion.

30. Hug. 26, fol. 72r; *O. c. Chr. Huygens,* 16:393.

31. "Ut ☐ EBD ad ☐ FBG ita BF ad BX [. . .]" (Hug. 26, fol. 72r; *O. c. Chr. Huygens,* 16:393).

32. Hug. 26, fol. 72r; *O. c. Chr. Huygens,* 16:393–394.

33. See nn. 17 and 25, above.

34. Galileo, *Two New Sciences,* 34 (*Opere,* 8:73).

35. On the early history of the Royal Academy, see the epilogue, below.

36. See *O. c. Chr. Huygens,* vol. 19, esp. pp. 25ff.

37. The full title is *Horologium Oscillatorium sive de motu pendulorum ad horologia aptato demonstrationes geometricae* (Paris, 1673), reprinted in *O. c. Chr. Huygens,* 18:69–368. This work is available in English as *Christiaan Huygens' "The Pendulum Clock, or, Geometrical Demonstrations concerning the Motion of Pendula as Applied to Clocks"* (hereafter *Pendulum Clock*), trans. Richard J. Blackwell (Ames: Iowa State University Press, 1986).

38. Christiaan Huygens, "Sur les Règles du mouvement dans la rencontre des corps," *Journal des Sçavans* (18 March 1669); see also *De motu corporum ex percussione* (Leyden, 1703), in *O. c. Chr. Huygens,* vol. 16.

39. The five parts of the work are entitled (1) Description of the Pendulum Clock; (2) Of the Fall of Heavy Bodies and of their Cycloidal Motion; (3) Of the Evolution and of the Dimension of Curved Lines; (4) Of the Center of Oscillation; (5) New Construction Based on the Circular Motion of Pendulums, and Theorems about the Centrifugal Force.

40. Huygens, *Pendulum Clock,* 33–34.

41. See Koyré, "Galilée et la loi d'inertie," in *Études galiléennes,* 161ff.

42. See Clavelin, *La philosophie naturelle de Galilée,* 373ff.

43. Galileo, *Two New Sciences,* 197 (*Opere,* 8:243). It is remarkable that Galileo is unable here to abstract from the underlying horizontal plane.

44. Ibid., 217 (*Opere,* 8:268).

45. Descartes, *Principles of Philosophy,* 1:241–242 (AT, 9 [§39]:85–86). For a study of the history of the principle of inertia and its various formulations, see Koyré, *Études galiléennes.*

46. Galileo, *Two New Sciences*, 217 (*Opere*, 8:268).
47. See above.
48. Huygens, *Pendulum Clock*, 34.
49. Ibid.
50. Ibid., 36. The passages excerpted in the preceding paragraph can be found at p. 35.
51. Ibid., 36.
52. Ibid.
53. Ibid., 36–37.
54. Ibid., 37.
55. Ibid.
56. Ibid., 38.
57. Ibid., 40.
58. Ibid., 40–42.
59. Ibid., 42.
60. Ibid., 69.

Chapter Two

1. Aristotle *Physics* 8.9.265a12–265b16, in *The Complete Works of Aristotle,* rev. Oxford trans., ed. Jonathan Barnes, Bollingen Series, no. 71 (Princeton: Princeton University Press, 1984), 1:442–443.

2. Nicholas Copernicus, *De revolutionibus orbium coelestium,* trans. John F. Dobson and Selig Brodetsky (London: Royal Astronomical Society, 1947), quoted with slight modifications in Thomas S. Kuhn, *The Copernican Revolution: Planetary Astronomy in the Development of Western Thought* (Cambridge: Harvard University Press, 1957), 147. (Kuhn replaces every instance of "orbit" by "sphere," as here, or by "circle.") Compare the more recent rendering of Edward Rosen, in *Nicholas Copernicus: On the Revolutions,* ed. Jerzy Dobrzycki (Baltimore: Johns Hopkins University Press, 1978), 10: "I shall now recall to mind that the motion of the heavenly bodies is circular, since the motion appropriate to a sphere is rotation in a circle. By this very act the sphere expresses its form as the simplest body, wherein neither beginning nor end can be found, nor can the one be distinguished from the other, while the sphere itself traverses the same points to return upon itself."

3. Johannes Kepler, *Epitome astronomiae Copernicae,* quoted in Alexandre Koyré, *The Astronomical Revolution: Copernicus, Kepler, Borelli,* trans. R. E. W. Watson (Ithaca: Cornell University Press, 1973), 289–90. On the expression of "driving force" (sometimes referred to as the "solar force"), see more specifically Kepler, *Gesammelte Werke,* ed. Walther van Dyck, Max Caspar, and F. Hammer (Munich: C. H. Beck, 1937–), 7:304ff.

4. Descartes, *Principles of Philosophy,* in *Philosophical Writings,* 1:259–260.

5. This text can be found in *O. c. Chr. Huygens,* 16:254–301, as well.

6. Ibid., 268.

7. Ibid., 269.

8. Descartes, AT, 2:380; quoted in François De Gandt, *Force and Geometry in Newton's "Principia,"* trans. Curtis Wilson (Princeton: Princeton University Press,

1995), 118. The term "gravity" in Wilson's rendering has been changed here to "weight."

9. Descartes, AT, 2:385; quoted in De Gandt, *Force and Geometry*, 119.

10. Descartes, *Le Monde, ou Traité de la lumière,* trans. Michael Sean Mahoney (New York: Abaris Books, 1979), 127–133 (AT, 11:74–77). The explanation given in the *Principia* rests on a very similar analysis: see AT, 9 (pt. 4, §§20–27): 210–214.

11. Descartes, *Principles of Philosophy,* 1:247 (AT, 9 [pt. 2, §64]: 101–102).

12. Various members presented papers on the subject on successive Wednesdays during the second half of the year: Roberval on 7 August 1669 (Registres manuscrits des procès-verbaux des séances de l'Académie Royale des Sciences de Paris [hereafter A. Ac. Sc. Registres], vol. 5, fols. 129r–132r), Frénicle on 14 August (ibid., fols. 133r–150v), Buot on 21 August (ibid., fols. 151r–158v), Huygens on 28 August (ibid., fols. 164r–179v), Du Hamel on 6 November (ibid., fols. 198r–205r), Mariotte on 13 November (ibid., fols. 206r–212r), and also Perrault (ibid., fols. 213r–222r). Note too the discussion between Roberval and Huygens on 4 September (ibid., fols. 180r–183v), continued on 23 October (ibid., fols. 191v–194r).

13. Ibid., fol. 129r.

14. Ibid., fol. 129r–129v.

15. Ibid., fol. 129v.

16. Ibid.

17. Ibid.

18. The reference here is to Archimedes' *On the Equilibrium of Planes.* The reader may consult either the Leipzig edition of the complete works, *Opera Omnia,* ed. J. L. Heiberg, 2d ed., 3 vols. (Leipzig: Teubner, 1910–1915); or, in English, *The Works of Archimedes,* ed. T. L. Heath (Cambridge: Cambridge University Press, 1897; reprint, New York: Dover, 1912, 1950).

19. A. Ac. Sc. Registres, vol. 5, fols. 131v–132r.

20. Ibid., fol. 129v.

21. Ibid., fol. 151r.

22. Ibid., fol. 152v.

23. Ibid., fol. 129v.

24. Ibid., fol. 141v.

25. Ibid., fol. 144r–144v.

26. Ibid., fol. 206r.

27. Ibid., fols. 207v–208v.

28. Ibid., fol. 129v.

29. This was also the position, although with certain qualifications, of Jean-Baptiste Du Hamel (1624–1706) and of Claude Perrault (1613–1688). For details concerning their papers, see n. 12, above.

30. A. Ac. Sc. Registres, vol. 5, fol. 164r–164v.

31. Ibid., fols. 164v–165r.

32. Ibid., fols. 166v–169r.

33. Ibid., fol. 172r–172v.

34. Ibid., fol. 178v.

35. Ibid., fol. 179r–179v.

36. Three editions of the *Principia* appeared during Newton's lifetime, in 1687, 1713, and 1726. The present study is based on the 1726 edition, first translated into English by Andrew Motte in 1729 and subsequently revised by Florian Cajori some two hundred years later: see *Sir Isaac Newton's Mathematical Principles of Natural Philosophy and His System of the World* (hereafter Cajori edition) (Berkeley and Los Angeles: University of California Press, 1934); a paperback edition in two volumes first appeared in 1966. More recently an annotated version of the 1726 Latin edition was published by Alexandre Koyré and I. Bernard Cohen, *Philosophiae Naturalis Principia Mathematica: The Third Edition (1726) with Variant Readings* (Cambridge: Cambridge University Press, 1972). A new edition of Motte-Cajori is forthcoming from the University of California Press, edited by I. Bernard Cohen, that aims to correct errors in the first edition and to modernize certain archaic mathematical terms and other language, in keeping with Cajori's ambitions for his own revision of the text, which remained incomplete at his death (see R. T. Crawford's note at p. ix of the Cajori edition). The standard French translation, revised by Alexis-Claude Clairaut, is by Gabrielle-Émilie de Breteuil, Marquise du Chastelet, *Principes mathématiques de la philosophie naturelle,* 2 vols. (Paris, 1756–1759), twice reprinted in recent years (Paris: Blanchard, 1966; Gabay, 1989). All passages cited in what follows are taken from the Cajori edition (1966 paperback ed.) unless otherwise noted.

37. For a detailed study of these questions, see Michel Blay, "Le traitement newtonien du mouvement des projectiles dans les milieux résistants," *Revue d'histoire des sciences* 40, nos. 3–4 (1987): 325–355.

38. See chap. 3, below.

39. Descartes, *René Descartes: Principles of Philosophy,* trans. V. R. Miller and R. P. Miller (Dordrecht: D. Reidel, 1983), 94.

40. Newton, *Principia,* 1.

41. Ibid. In the commentary that follows this first definition, Newton adds, "It is this quantity that I mean hereafter everywhere under the name of body or mass." By contrast there is no definition of velocity (see chap. 3, below).

42. Ibid., 2. Definition III concerns the *vis insita,* the internal or innate force that resides in matter, that is, the "power of resisting, by which every body, as much as in it lies, continues in its present state, whether it be of rest, or of moving uniformly forwards in a right line" (ibid.).

43. Ibid.

44. Ibid., 6.

45. Ibid., 13.

46. Ibid.

47. Ibid.

48. See *The Mathematical Papers of Isaac Newton,* ed. D. T. Whiteside (Cambridge: Cambridge University Press, 1974), 6:30ff. Vol. 6 covers the years 1684–1691.

49. Newton did, however, introduce in lemma II of book II a quasi-algorithmic procedure, announcing his calculus of fluxions, in order to resolve tricky problems of ballistics that required taking into account the variation of a squared term. See n. 37, above.

50. Newton, *Principia,* 38.

51. Lazare Carnot, *Reflexions on the Metaphysical Principles of the Infinitesimal Analysis,* trans. W. R. Browell (Oxford, 1832), 86–87. The original work, published in Paris in 1797, was reissued in a second edition in 1813, and subsequently in two volumes by Gauthier-Villars in 1921. On the Newtonian method, see too François De Gandt, "Le style mathématique des *Principia* de Newton," *Revue d'histoire des sciences* 39, no. 3 (1986): 195–222.

52. Newton, *Principia,* 29.

53. Ibid., 32.

54. Lemma VI reads in full as follows: "If any arc ACB, given in position, is subtended by its chord AB, and in any point A, in the middle of the continued curvature, is touched by a right line AD, produced both ways; then if the points A and B approach one another and meet, I say, the angle BAD, contained between the chord and the tangent, will be diminished *in infinitum,* and ultimately will vanish" (ibid.).

55. Ibid., 32–33.

56. Ibid., 40.

57. On this topic see Michel Blay and Georges Barthélemy, "Changements de repères chez Newton: Le problème des deux corps dans les *Principia,*" *Archives internationales d'histoire des sciences* 34, no. 112 (1984): 69–98.

58. Newton, *Principia,* 40.

59. Ibid.

60. See Blay, "Le traitement newtonien du mouvement des projectiles dans les milieux résistants."

61. Newton, *Principia,* 236.

62. Ibid., 41.

63. Ibid.

64. Ibid.

65. Ibid.

66. "And therefore these ultimate figures (as to their perimeters *ac*E) are not rectilinear, but curvilinear limits of rectilinear figures" (ibid., 30).

67. Ibid., 41.

68. Ibid., 41–42.

69. "A body, acted on by two forces simultaneously, will describe the diagonal of a parallelogram in the same time as it would describe the sides by those forces separately" (ibid., 14).

70. Ibid., 42.

71. Ibid., 48.

72. Ibid., 36.

73. Ibid., 37.

74. Ibid.

75. See below.

76. Ibid. The second and third corollaries of lemma XI do not figure in the edition of 1687, from which it may be inferred that certain modifications to proposition VI were made, by comparison with the original edition, in the course of preparing the 1713 and 1726 editions; on this point see Koyré and Cohen, eds., *Philosophia Naturalis Principia Mathematica,* 102–107.

77. A detailed study of this proposition has been made by Georges Barthélemy

in his unpublished 1985 Sorbonne doctoral thesis, "Concepts et méthodes de la mécanique rationelle dans les *Principia* de Newton."

78. Newton, *Principia*, 48.

79. Ibid., 48–49.

80. Ibid., 49.

81. Ibid., 50.

82. Ibid.

Chapter Three

1. The original text of this work is to be found in Galileo, *Opere*, 8:43–313.

2. See chap. 1, above.

3. See chap. 2, above.

4. Galileo, *Two New Sciences*, 148–149.

5. René Descartes, *Regulae ad directionem ingenii* (Leyden, 1629), in AT, 10:359–469; citations to this work in what follows are taken from the English translation by J. Cottingham et al., *Rules for the Direction of the Mind*, in *Philosophical Writings*, 1:7–78.

6. René Descartes, *La géométrie*, published as part of *Discours de la méthode: pour bien conduire sa raison, & chercher la vérité dans les sciences; Plus La dioptrique; Les météores; Et La géométrie. Qui sont des essais de cette méthode* (Leyden, 1637), in AT, 6:369–485; see the translation by David Smith and Marcia Latham, *The "Geometry" of René Descartes* (New York: Dover, 1954).

7. Descartes, *Rules*, 68 (AT, 10:456–457). The reference to "the arbitrary unit mentioned above" alludes to this prior passage: "Unity is the common nature which [. . .] all the things which we are comparing must participate in equally. If no determinate unit is specified in the problem, we may adopt as unit either one of the magnitudes already given or any other magnitude, and this will be the common measure of all the others. We shall regard it as having as many dimensions as the extreme terms which are to be compared" (*Rules*, 63–64).

8. Galileo, *Two New Sciences*, 148 (*Opere*, 8:191).

9. Ibid., 165 (*Opere*, 8:208).

10. Ibid.

11. Certain translations (such as that by Maurice Clavelin) render the preceding clause as "[. . .] the sum of all parallels contained in the quadrilateral equal to the sum of those included in triangle AEB," but "aggregate" is a more faithful rendering of Galileo's original term, for which reason it is to be preferred. The full Latin text is as follows: "Quod si parallelae trianguli AEB usque ad IG extendantur, habemus aggregatum parallelarum omnium in quadrilatero contentarum aequalem aggregatui comprehensarum in triangulo AEB" (*Opere*, 8:208–209). The term "aggregate" echoes the work of Cavalieri (1598–1647); see below.

12. It should be noticed that Galileo here introduces the expression "momenta of speed" in place of "degrees of speed." Even if these two expressions are for the most part synonymous in Galileo's writings, the substitution of the one for the other in this particular part of the argument nonetheless remains somewhat problematic.

13. Galileo, *Two New Sciences*, 165–166 (*Opere*, 8:208–209).

14. Ibid., 165 (*Opere*, 8:208).

15. If one wishes to remain strictly faithful to the Galilean conceptualization, without modernizing it, one must be careful not to interpret this mathematical relation as obtaining between each degree of velocity and (by virtue of the properties of uniform motion) each elementary space traversed in an instant.

16. On this difficulty with Galileo's argument see Clavelin, *La philosophie naturelle de Galilée*, 309–310; Jacques Merleau-Ponty, *Leçons sur la genèse des théories physiques: Galilée, Ampère, Einstein* (Paris: Vrin, 1974), 50–51; and Koyré, *Études galiléennes*, 149ff. One may also consult Paolo Galluzzi, *Momento: Studi galileiani* (Rome: Edizione dell'Ateneo and Bizzarri, 1979); Enrico Giusti, "Aspetti matematici della cinematica galileiana," *Bollettino di storia della scienze matematiche 5*, no. 2 (1981): 3–42 (esp. 32); and Antonio Nardi, "La quadratura della velocitè: Galileo, Mersenne, La Tradizione," *Nuncius* 3 (1988): 27–64.

17. Galileo, *Two New Sciences,* 166–167 (*Opere*, 8:209–210).

18. Torricelli, *Opere,* 1 (pt. 2):259; cited in De Gandt, *Force and Geometry,* 113 (N.B.: The lettering has been altered in the quoted text to reflect Torricelli's original figure, which was modified by De Gandt).

19. In this connection see particularly François De Gandt, "Les indivisibles de Torricelli," in his edited volume *L'oeuvre de Torricelli: Science galiléenne et nouvelle géométrie* (Nice: Presses de l'Université de Nice, 1989), 151–206 (esp. 194–197); and, in this same volume, Ettore Bortolotti's article, translated by P. Souffrin and J.-P. Weiss, "L'oeuvre géométrique d'Évangeliste Torricelli," 115–146 (esp. 132–135).

20. A. Ac. Sc. Registres, vol. 13, fols. 76r–77v. See also *Histoire de l'Académie Royale des Sciences* (Paris, 1733) (henceforth *Histoire*), 2:155–157; note that the first volume of this work covers the years 1666–1686 ("Depuis son établissement en 1666 jusqu'à 1686"), and the second, from 1686 to 1699 *(Depuis 1686 jusqu'à son renouvellement en 1699)*.

21. A. Ac. Sc. Registres, vol. 13, fol. 76r. See also *Histoire*, 2:155, where the last sentence reads: "M. Varignon shows how he could have done it, even following his own principles."

22. See Galileo, *Opere*, 8:255. In what follows, citation is to the English translation by Stillman Drake, *Dialogue concerning the Two Chief World Systems— Ptolemaic and Copernican* (Berkeley and Los Angeles: University of California Press, 1953).

23. Ibid., 228.

24. The analysis of the beginning of the motion is a very tricky problem that will be developed in the following paragraph. In this connection it is interesting to note the remark made by Cavalieri in a letter to Galileo dated 19 December 1634: "Now, since the beginning and end of a motion are not motion [. . .]" (*Opere*, 16:174).

25. Galileo, *Dialogue,* 228–229.

26. Ibid., 229.

27. Ibid.

28. Ibid.

29. Ibid.

30. Ibid.

31. See n. 78 to this chapter, below.

32. Galileo, *Dialogue*, 229.

33. The figure reproduced in the paper presented to the Royal Academy in 1692 contains an additional parallel line, HK, between FG and MN; the figure reproduced here is from the second volume of the *Histoire* (1733), p. 156.

34. A. Ac. Sc. Registres, vol. 13, fol. 76v. The figure accompanying this passage is taken from the *Histoire*, 2:156, the text of which slightly departs from the one cited here: "Let AB be any line whatever that expresses whatever time one likes of the fall of a body, since by hypothesis the speeds of this body [. . .]"; and "[. . .] and thus in all the other imaginable parts of the time AB until BC, which will express the whole speed of this body at the end of all this time [. . .]."

35. See the next section of this chapter.

36. A. Ac. Sc. Registres, vol. 13, fol. 77r; *Histoire*, 2:156.

37. Ibid.

38. The corresponding sentence in the *Histoire* (2:156) speaks of "infinitely near" lines.

39. A. Ac. Sc. Registres, vol. 13, fol. 77r.

40. Ibid.

41. Ibid.

42. Ibid., fol. 77r–77v.

43. The words "sum of," left out originally, were entered between the lines of the manuscript by the same hand. The *Histoire* (2:157) reads: "[. . .] in falling are as the sums of the speeds [. . .]."

44. A. Ac. Sc. Registres, vol. 13, fol. 77r.

45. The original Latin is "Effectus sunt causis suis adaequatus proportionales"; see John Wallis, *Mechanica sive de motu tractatus geometricus* (London, 1670–1671), in *Opera Mathematica* (Oxford, 1695–1699), 1:584.

46. "Universalem hanc Propositionem praemittendam etiam duxi, quoniam viam aperit qua, ex pure Mathematica speculatione, ad Physicam transeatur; seu potius hanc et illam connectit" (ibid., 1:584).

47. On this question see Michel Blay, "Varignon et le statut de la loi de Torricelli," *Archives internationales d'histoire des sciences* 35, nos. 114–115 (1985): 330–345.

48. See Torricelli, *De motu gravium*, particularly pp. 191–204 in the second book, of which the *De motu aquarum* forms a part (*Opere*, 2:185ff.).

49. "Aquas violenter erumpentes in ipso eruptionis puncto eundem impetum habere, quem haberet grave aliquod, sive ipsius aquae gutta una, si ex suprema eiusdem aquae superficie usque ad orificium euptionis naturaliter cecidesset" (ibid., 191).

50. "Experimentum etiam aliquo modo principium nostrum probat, quamquam aliqua ex parte reprobare videatur" (ibid., 192).

51. A. Ac. Sc. Registres, vol. 3, fol. 99v. For a detailed analysis of this question, see in particular Michel Blay, "Recherches sur les forces exercées par les fluides en mouvement à l'Académie Royale des Sciences: 1668–1669," in *Mariotte, savant et philosophe: Analyse d'une renommée*, ed. Pierre Costabel and Michel Blay (Paris: Vrin, 1986), 91–124.

52. A. Ac. Sc. Registres, vol. 3, fol. 99r.

53. Ibid., fol. 113r; *O. c. Chr. Huygens*, 19:171.

54. A. Ac. Sc. Registres, vol. 3, fol. 113r–113v.

55. Ibid., fols. 113v–117v; *O. c. Chr. Huygens*, 19:166–170.

56. A. Ac. Sc. Registres, vol. 14, fol. 92r.

57. Ibid., fols. 94r–96v.

58. Ibid., fol. 94r–94v. This paper was reprinted with slight modifications in *Histoire*, 2:260–262.

59. A. Ac. Sc. Registres, vol. 14, fol. 94v.

60. Ibid., fol. 95r. A very interesting analysis by Fontenelle of Varignon's approach is found in the volume of *AM* for the year 1703 (1705): 125.

61. This qualification is not explicit in Varignon's text of 1695, but it was made so when the demonstration was reprinted posthumously thirty years later; see *Traité du mouvement et de la mesure des eaux coulantes et jalissantes* (Paris, 1725), 51–52.

62. A. Ac. Sc. Registres, vol. 14, fol. 95r.

63. Pierre Varignon, "Règles du mouvement en général," in *Mémoires de l'Académie Royale des Sciences depuis 1666 jusqu'à 1699* (Paris: 1730), 10:225–233. Note that the nine volumes that complete the series of *Mémoires* are numbered 3–11 since the first two volumes, not published until three years later, in 1733, are constituted by the *Histoire* of the Royal Academy.

64. A. Ac. Sc. Registres, vol. 14, fol. 96r–96v.

65. Pierre Varignon, *Nouvelle mécanique ou statique dont le projet fut donné en 1687: Ouvrage posthume de M. Varignon* (Paris, 1725), 4.

66. Varignon, *Traité du mouvement*, 2.

67. J.-F. Montucla, *Histoire des mathématiques*, 2d ed., 4 vols. (Paris, 1799–1802), 3:682; see also Joseph-Louis Lagrange, *Mécanique analytique*, 3d ed. (Paris, 1853–1855), 2:245–246.

68. Pierre Varignon, "Du mouvement des eaux, ou d'autres liqueurs quelconques de pesanteurs spécifiques à discrétion; de leurs vitesses, de leurs dépenses par telles ouvertures ou sections qu'on voudra; de leurs hauteurs au-dessus de ces ouvertures, des durées de leurs écoulemens etc.," *AM*, year 1703 (1705): 238–261.

69. Ibid., 261.

70. "Definition XIII: The weights of bodies equal in volume are called specific gravities, or gravities peculiar to the kinds of these bodies." See Varignon, *Nouvelle mécanique*, 61 n. 66.

71. "Definition XI: The magnitude of a body, or otherwise the space that it occupies, is called the volume of this body. The combination or linkage of the particles of which the body is composed is called its density. And the totality, or the sum, of all these particles is called the mass" (ibid., 55).

72. *AM*, year 1703 (1705): 238.

73. Ibid., 250.

74. Ibid., 240. The Newtonian notation was apparently introduced only at the printing stage, for the manuscript uses the notations dm and $\delta\mu$ (A. Ac. Sc. Registres, vol. 22, fol. 325v). It was therefore evidently used here only for the sake of convenience.

75. *AM*, year 1703 (1705): 245.

76. See A. Ac. Sc. Registres, vol. 22, fol. 326v. The missing part of the original text of the minutes was rewritten at the time of publication.

77. The following statement occurs in Varignon's previously mentioned paper "Règles du mouvement en général," which he read to the academy on 31 December 1692: "[. . .] M. Wallis commenced his Mechanics by a Treatise on Motion in general; but the path that he took yet led him only to very few Rules; besides, he proves them all only by induction, and never in a general and universal manner" (*Mémoires de l'Académie Royale des Sciences depuis 1666 jusqu'à 1699*, 10:225).

78. *AH*, year 1703 (1705): 126. The lines quoted are taken from the lecture notice of 1703 but refer more directly to the 1695 paper.

79. Galileo, *Opere*, 8:312.

80. Pascal, *Great Shorter Works*, 195. *De l'esprit géométrique*, left unpublished at the time of Pascal's death, is believed to have been written during the period 1657–1658.

81. Galileo, *Two New Sciences*, 157.

82. Ibid.

83. On this point see the commentary of Léon Brunschvicg, *Les étapes de la philosophie mathématique* (Paris: Blanchard, 1972), 215.

84. In this connection see Hermann Weyl, *Philosophy of Mathematics and Natural Science* (Princeton: Princeton University Press, 1949), 160–161.

85. Gottfried Wilhelm Leibniz, *Philosophical Papers and Letters*, trans. and ed. Leroy E. Loemker, 2d ed. (Dordrecht: D. Reidel, 1976), 544 (GM, 4:93).

86. Gottfried Wilhelm Leibniz, *Die Philosophischen Schriften von Leibniz*, ed. C. I. Gerhardt, 7 vols. (Hildesheim: Georg Olms, 1960–1961) (hereafter GP), 2:104–105. The full text of the letter, dated 22 July/1 August 1687, can also be found in *The Leibniz-Arnauld Correspondence*, ed. and trans. H. T. Mason (Manchester: Manchester University Press, 1967), 130–131.

87. Gottfried Wilhelm Leibniz, *New Essays on Human Understanding*, trans. and ed. Peter Remnant and Jonathan Bennett (Cambridge: Cambridge University Press, 1981); the pages of the preface, in which this passage occurs, are unnumbered. See also the letters from Leibniz to Volder dated 1688 and 24 March/3 April 1689 (GP, 2:161 and 168–169, respectively); to Hartsoeker, dated 30 October 1710 (GP, 3:506); and to Jean Bernoulli, dated 20/30 September 1698 (GM, 3:544); and also the "Essay de Dynamique sur les lois du mouvement" of 1699 (esp. GM, 6:229) and the "Specimen Dynamicum" of 1695 (esp. GM, 6:249).

88. Gottfried Wilhelm Leibniz, "Letter of Mr. Leibniz on a General Principle Useful in Explaining the Laws of Nature through a Consideration of the Divine Wisdom; to Serve as a Reply to the Response of the Rev. Father Malebranche" (July 1687), in *Philosophical Papers and Letters*, 351–354. From a modern mathematical point of view, as Loemker remarks in a note to the text, the law of continuity is interpreted to stipulate that "if $y = f(x)$, and there are two values x_1 and x_2 such that $x_2 - x_1 < d$, where d is any assignable difference, however small, then the corresponding values $y_2 - y_1 <$ any assignable difference as well" (ibid., 354). The original French text can be found in GP, 3:51–55 (see esp. 52); see also the Latin manuscript in GM, 6:129–135 (pt. 6); and, for commentary on this passage, André Robinet, *Malebranche et Leibniz: Relations personelles présentées*

avec les textes complets des auteurs et de leurs correspondants revus, corrigés et inédits (Paris: Vrin, 1955), 258–264.

89. Gottfried Wilhelm Leibniz, *De Corporum Concursu*, January–February 1678, Leibniz-Handschriften, Mathematik, vol. 35, 9, fol. 23, Niedersächsische Landesbibliothek, Hanover. I am grateful to Michel Fichant for calling my attention to this manuscript.

90. Gottfried Wilhelm Leibniz, *Sämtliche Schriften und Briefe*, 2d ser., 1st volume (Berlin: Akademie Verlag, 1926), 470.

91. Descartes, AT, 2:399.

92. See Descartes, *Philosophical Writings*, 1:244–245 (AT, 9 [pt. 2, §§45–52]), especially the note to §45 in the Cambridge edition summarizing "Descartes' seven rules for calculating the speed and direction of bodies after impact"; the full text of these rules, omitted there, can be located in *Descartes, Principles of Philosophy*, 64ff.

93. See in this connection Leibniz's *De Corporum Concursu* (n. 89, above) as well as his letters to Malebranche, one undated and the other dated 8/12 October 1698, in GP, 1:350 and 354, respectively. For Leibniz, the law of continuity had, on the one hand, an antiatomistic dimension, as he noted in his letter of January 1692 to Foucher: "My Axiom that nature never acts by leap[s], which you [de]-mand that R. P. Malebranche approve, is of the greatest usefulness in physics; it destroys atomos, quietulas, globulos, secundi Elementi and other similar chimeras [. . .]" (GP, 1:403); and, on the other hand, it led to a certain analogy between the microscopic and the macroscopic, as Leibniz's letters to Varignon of 2 February 1702 (GP, 4:93–94), to Wolf of 9 November 1705 (in *Briefwechsel zwischen Leibniz und Christian Wolf aus den Handschriften der Koeniglichen Bibliothek zu Hannover*, ed. C. I. Gerhardt [Halle, 1860], 44), and to Hartsoeker in 1710 (GP, 2:500) indicate.

94. Descartes, *Principles of Philosophy*, 64–65 (AT, 9 [pt. 2, §46]).

95. Ibid., 65 (AT, 9 [pt. 2, §47]).

96. Ibid., 65 (AT, 9 [pt. 2, §48]).

97. Leibniz, *Philosophical Papers and Letters*, 399 (GP, 4:377).

98. Ibid., 397 (GP, 4:375).

99. Ibid.

100. Ibid. See also Leibniz's "Tentamen Anagogicum: Essay anagogique dans la recherche des causes," in GP, 7:279.

101. Leibniz, *Philosophical Papers and Letters*, 400 (GP, 4:378).

102. François Bernier, *Abrégé de la philosophie de Gassendi* (Lyons, 1678), 1:296–299; the 1684 edition (in seven volumes rather than the original eight) contains a much more nuanced statement of Bernier's position (see 2:191ff.). See too Pierre Gassendi, *Syntagma Philosophicum*, in *Opera Omnia in sex tomos divisa* (Lyons, 1658), 1:341ff.

103. Edme Mariotte, *Traitté de la percussion ou chocq des corps: Dans lequel les principales Règles du mouvement contraires à celles de Mr. Descartes, & quelques autres Modernes ont voulu établir, sont démonstrées par leurs véritables causes* (Paris, 1673). This work was to be republished many times, Descartes's name eventually disappearing from the title. It was also reprinted in the *Oeuvres de Mr. Mariotte* (Leyden, 1717), 1:3–116; see esp. 80–81.

104. Mariotte, *Traitté de la percussion ou chocq des corps*, 247–249 (note placed at the end of proposition X).

105. A. Ac. Sc. Registres, vol. 7, fol. 118v.

106. Leibniz, GP, 3:531.

107. Ibid., 3:533–534.

108. In his *Harmonie universelle contenant la théorie et la pratique de la musique* (Paris, 1636), Père Marin Mersenne, in proposition II of the second book ("Des mouvemens de toutes sortes de corps"), writes: "But because this speed continuously increases from moment to moment, and not by pauses, or leaps, from [one] time to [another], it is certain that the degrees of speed from the [point of rest] A until the acquisition of the degree HD in the time AD are infinite, in keeping with the infinity of instants of the time AD, or of the points of the line AD" (p. 89).

109. Thomas Hobbes, *Elementorum philosophiae sectio prima De Corpore* (London, 1655). To the criticisms formulated by John Wallis in his *Elenchus geometriae Hobbianae sive, geometricorum, quae in ipsius elementis philosophiae, a Thoma Hobbes Malmesburiensi proferuntur* (Oxford, 1655), Hobbes replied by publishing the following year an English version of *De Corpore* (in which "*conatus*" is translated as "endeavor"), appending to it a new work, *Six Lessons to the Professors of Mathematicks of the Institution of Sir Henry Savile, in the University of Oxford*. This quarrel was to go on for twenty-five years.

110. Thomas Hobbes, *Elements of Philosophy, the first section: concerning body* (London, 1656), in *The English Works of Thomas Hobbes of Malmesbury*, ed. William Molesworth (London: 1839–1845), 1 (pt. 2, chap. 8, §10):109.

111. Diogenes Laertius *Vitae philosophorum* 9.72.

112. Cited by Maurice Caveing in his *Zénon d'Élée, Prolégomènes aux doctrines du continu: Étude historique et critique des fragments et témoignages* (Paris: Vrin, 1982), 63–64.

113. Hobbes, *Elements of Philosophy*, in *English Works*, 1 (pt. 2, chap. 8, §11):110–111.

114. Ibid., 1 (pt. 3, chap. 15, §1):204.

115. Ibid., 1 (pt. 3, chap. 15, §2):206.

116. Ibid.

117. Ibid.

118. Euclid, *The Thirteen Books of Euclid's "Elements,"* trans. Thomas L. Heath, 2d rev. ed. (New York: Dover, 1956), 1:153.

119. Hobbes, *The Six Lessons to the Professors of Mathematicks*, in *English Works*, 7:201.

120. Thomas Hobbes, *De Principiis et Ratio-cinatione geometrarum* (London, 1666), in *Opera Philosophica*, ed. William Molesworth (London, 1839–1845), 4:392. Against this view, Roberval was to write in his unfinished *Élémens de géométrie* (on which he was still working at the time of his death in 1675): "The extremity of a finite and terminated line is called a point and has no extension. Thus it is indivisible, having neither position nor size." (The first published edition of this work was issued by Éditions Vrin in 1996; the passage quoted is from the manuscript copy held in the archives of the Academy of Sciences in Paris.)

121. Hobbes, *Elements of Philosophy,* in *English Works,* 1 (pt. 3, chap. 15, §2):206.

122. Ibid., §7:216–217.

123. The full title of the *New Physical Hypothesis* of 1670–1671 is *Hypothesis physica nova qua phaenomenorum naturae pleororumque causae ab unico quodam universali motu, in globo nostro supposito, neque tychonicis, neque copernicanis aspernando, repetuntur* (Mayence, 1671). The original text has been most recently reprinted in *Sämtliche Schriften und Briefe* (Berlin: Akademie Verlag, 1966), ser. 6, vol. 2, 262–334; and earlier in GM, 6:17–80. For reasons of convenience, reference will be made in what follows to GM.

124. Gottfried Wilhelm Leibniz, *Theoria motus concreti, seu hypothesis de rationibus phaenomenorum nostri orbis,* in GM, 6:18–59. This work is dedicated to the Royal Society of London (see ibid., 18).

125. Gottfried Wilhelm Leibniz, *Theoria motus abstracti seu rationes motuum universales, a sensu et phaenomenis independentes,* in ibid., 61–80. This work is dedicated to the Royal Academy of Sciences in Paris (see ibid., 62). A partial English translation can be found in Loemker's edition of the *Philosophical Papers and Letters,* 139–142.

126. See Leibniz's 1693 letter to Foucher in GP, 1:415.

127. Thus per the seventh principle: *Motus est continuus seu nullis quietulis interruptus* (GM, 6:68). Leibniz continues: "8. For where a thing is once at rest, it will always remain at rest unless a new cause of motion occurs. 9. Conversely, a thing once moved will always move with the same velocity [. . .]" (*Philosophical Papers and Writings,* 140).

128. See Bernier, *Abrégé de la philosophie de Gassendi.*

129. Leibniz, *Philosophical Papers and Writings,* 139. The Latin statement of these three principles is as follows: "1. *Dantur actu partes in continuo [. . .],*" "2. *Eaeque infinitae actu [. . .],*" and "3. *Nullum est minimum in spacio aut corpore [. . .]*" (GM, 6:67).

130. Ibid. (GM, 6:67–68).

131. Ibid. ("4. *Dantur indivisibilia seu inextensa [. . .]*"; GM, 6:68).

132. Ibid., 139–40 (GM, 6:68).

133. Pierre Costabel, *Leibniz and Dynamics: The Texts of 1692,* trans. R. E. W. Maddison (Ithaca: Cornell University Press, 1973), 17.

134. Leibniz, *Philosophical Papers and Writings,* 140 (GM, 6:67).

135. Ibid., 68. See too Leibniz's letter of 28 September 1670 to Oldenburg, in *Sämtliche Schriften und Brief* (1926), ser. 2, vol. 1:64; and the rough drafts of the *Theoria motus abstracti,* particularly GM, 6 (pt. 2):171.

136. Leibniz, *Philosophical Papers and Writings,* 140 (GM, 6:68).

137. Foucher had already noted in his letter to Leibniz of 31 December 1691: "He [Malebranche] says that he is of the same sentiment as you regarding nature's manner of acting by infinitely small changes and never by leap[s]. [As] for me, I confess to you that I still have my doubts, for I believe that this comes back to the argument of the Pyrrhonians, who made the tortoise march as fast as Achilles; for all magnitudes being able to be divided to infinity, there is no point so small in which one cannot conceive an infinity of divisions that will never be exhausted. Whence it follows that these motions must occur at once, by comparison with

certain physical (and not mathematial) indivisibles. If you can break through the barrier that [stands] between physics and metaphysics by your problem, as you have thought, I would be most grateful to you for it; for the more uniformity that can be found in objects is [for] the best" (Leibniz, GP, 1:400).

138. Ibid., 411–412.

139. Leibniz, *Philosophical Papers and Writings,* 141 (GM, 6:70).

140. Ibid.

141. Leibniz, GP, 1:412.

142. Grégoire de Saint-Vincent, *Opus geometricum quadraturae circuli et sectioni coni* (Antwerp, 1647).

143. Leibniz, GP, 1:415–416.

Chapter Four

1. On this point see Joseph E. Hofmann, *Leibniz in Paris, 1672–1676: His Growth to Mathematical Maturity* (New York: Cambridge University Press, 1974), 187–201; J. M. Child, *The Early Mathematical Manuscripts of Leibniz, Translated from the Latin Texts* (Chicago: Open Court, 1920); Margaret E. Baron, *The Origins of the Infinitesimal Calculus* (New York: Dover, 1987), 253–290; Carl B. Boyer, *The History of the Calculus and Its Historical Development* (New York: Dover, 1959), 187–223; C. H. Edwards, Jr., *The Historical Development of the Calculus* (New York: Springer-Verlag, 1979), 231–267.

2. Publication of *Acta Eruditorum* (hereafter *AE*), in Leipzig, under the direction of Otto Mencke, commenced at the beginning of 1682.

3. Gottfried Wilhelm Leibniz, "Nova methodus pro maximis et minimis, itemque tangentibus, quae nec fractas, nec irrationales quantitates moratur, et singulare pro illis calculi genus," *AE* (October 1684): 467–473; later reprinted with modifications in Leibniz, GM, 5:220–226. The English title quoted, long standard, is cited (for example) in Boyer, *History of the Calculus,* 207. A slightly different, and more complete, rendering of the original Latin title is given by H. J. M. Bos: "A new method for maxima and minima, and also for tangents, which is not impeded by fractions or irrational quantities, and a singular kind of calculus for these" ("Differentials, Higher-Order Differentials and the Derivative in the Leibnizian Calculus," *Archive for History of Exact Sciences* 14, no. 1 [1974]: 29); in certain versions the last clause of the title reads "and a remarkable type of calculus for this." Regarding the original publication of the article and various early drafts, see Heinz-Jürgen Hess, "Zur Vorgeschichte der 'Nova Methodus' (1676–1684)," *Studia Leibnitiana,* Special Issue 14 (1986): 64–102. With regard to the Leibnizian calculus in its fully developed form, the reader is urged to consult the remarkable article by Bos cited above as well as his later article "Fundamental Concepts of the Leibnizian Calculus," *Studia Leibnitiana,* Special Issue 14 (1986): 103–118.

4. Leibniz, GM, 5:220. Note that the text in GM is a corrected version of the original article, and that in the corrected version Leibniz goes on immediately to treat differences as infinitesimals.

5. Since both the "indeterminate letter and its differential" play the same role in the calculus, it thus becomes possible to reiterate the operation.

6. Leibniz, GM, 5:220. Leibniz, as we would say today, contemplates only positive abcissas and ordinates.

7. Ibid., 221. Note that the order of "convexity" and "concavity" here is inverted in the original article (compare the translation in D. J. Struik, ed., *A Source Book in Mathematics, 1200–1800* [Princeton: Princeton University Press, 1986], 274–275). On the formulas involved in calculating the radius of curvature, see Jacques Bernoulli, "Curvatura veli," *AE* (May 1692): 202–211; "Curvae diacausticae," *AE* (June 1693): 244–256; and "Curvatura laminae elasticae," *AE* (June 1694): 262–276.

8. Leibniz, GM, 5:222.

9. Ibid., 5:222–223.

10. Here Leibniz recommends considering dx, dy, dv, dw, and dz as proportional to the differences, incremental or decremental, of x, y, v, w, and z. Ibid., 5:223.

11. Leibniz uses "member" here, whereas today we would say "term."

12. Ibid.

13. Ibid., 223–226.

14. Optics had already been the object of Leibniz's attention in a paper published four years earlier entitled "Unicum opticae, catoptricae et dioptricae principium," *AE* (June 1682): 185–190.

15. Gottfried Wilhelm Leibniz, "De geometria recondita et analysi indivisibilium atque infinitorum," *AE* (June 1686): 292–300; GM, 5:226–233.

16. John Craig, *Methodus figurarum lineis rectis et curvis comprehensarum quadraturas determinandi* (London, 1685).

17. Jacques Bernoulli, "Analysis problematis antehac propositi: De Inventione Lineae descensus a corpore gravi percurrendae uniformiter, sic ut temporibus aequalibus aequales altitudines emetiatur; et alterius cujusdam problematis Proposito," *AE* (May 1690): 217–220. The use of the terms "integral" and "integral calculus" was the subject of a priority dispute between Jacques and Jean Bernoulli. The term "integral" appeared for the first time in the article just cited by Jacques Bernoulli at page 218.

18. Here Leibniz means to emphasize that his method may be extended to handle nonalgebraic quantities.

19. This expression, introduced by François Viète (1540–1603) in his *Isagoge in artem analyticem* (Tours, 1591), refers to the art of writing equations with letter coefficients.

20. The reference is to a theorem given by Isaac Barrow in his *Lectiones geometricae* (Cambridge, 1670), lesson XI.

21. Leibniz, GM, 6:230–231.

22. See n. 17, above.

23. Gottfried Wilhelm Leibniz, "Réponse de M. L. à la Remarque de M. l'Abbé D.C. contenue dans l'article 1. de ces Nouvelles, mois de juin 1687, où il prétend soutenir une loi de la Nature avancée par M. Descartes," *Nouvelles de la République des Lettres* (September 1687): 952–956; and GP, 3:49–51. The problem of the isochronic curve is formulated thus: "To find a line of descent, in which the heavy body descends uniformly, and equably approaches the horizon in equal

times, The Analysis of the Cartesian Gentlemen [*Messieurs les cartésiens*] will perhaps give it easily."

24. To appreciate the diversity of the subjects treated by Jacques and Jean Bernoulli, one need only consult the impressively large collected works of these authors: *Jacobi Bernoulli, Basileensis, Opera,* 2 vols. (Brussels: Culture et Civilisation, 1967); *Johannis Bernoulli [. . .], Opera Omnia tam antea sparsim edita, quam hactenus inedita,* 4 vols. (Hildesheim: Georg Olms, 1968).

25. The lessons themselves have been published. See Paul Schafheitlin, "Johann Bernoulli lectiones de calculo differentialum," *Verhandlungen der Naturforschenden Gesellschaft* 34 (1922): 1–31; see also Jean Bernoulli, *Opera Omnia,* 3:387–558.

26. Guillaume de l'Hospital, *Analyse des infiniment petits pour l'intelligence des lignes courbes* (Paris, 1696). Varignon announced the publication of de l'Hospital's book to Jean Bernoulli in a letter dated 18 June 1696: see *Der Briefwechsel von Johann I Bernoulli* (hereafter *Briefwechsel*), ed. P. Costabel and J. Peiffer (Basel: Birkhäuser Verlag, 1988), 2:103.

27. See the analysis of this matter in my *La naissance de la mécanique analytique* (Paris: Presses Universitaires de France, 1992).

28. In addition to Varignon and Fontenelle, this group included Charles René Reyneau (1656–1728), Claude Jacquemet (1651–1729), Louis Raphaël Lévy Byzance (1646–1722), Bernard Lamy (1640–1715), Louis Carré (1663–1711), Pierre Rémond de Montmort (1678–1719), Joseph Sauveur (1653–1716), Joseph Saurin (1655–1737), [?] Guisnée (?–1718), Bernard Renaud d'Élisagaray (1652–1719).

29. André Robinet, "Le group malebranchiste introducteur du calcul infinitésimal en France," *Revue d'histoire des sciences* 13, no. 3 (1960): 287–308; see too Pierre Costabel, "Pierre Varignon (1654–1722) et la diffusion en France du calcul différentiel et intégral," *Conférences du Palais de la Découverte,* ser. D, no. 108 (4 December 1965): 1–28. Otherwise, a great deal of information relating to this topic can be found in volumes 17/2, 19, and 20 of the *Oeuvres complètes de Malebranche,* ed. André Robinet, 2d ed. (Paris: Vrin, 1967–1978).

30. The official minutes of the Wednesday session of the Royal Academy of Sciences on 17 June 1693 record that "Mr. Abbé Bignon proposed on behalf of Monseigneur de Pontchartrain Monsieur le Marquis de l'Hospital for membership in the Company" (A. Ac. Sc. Registres, vol. 13, fol. 135v).

31. A detailed analysis of this opposition can be found in my "Deux moments de la critique du calcul infinitésimal: Michel Rolle et George Berkeley," *Revue d'histoire des sciences* 39 (1986): 223–253; and in my *La naissance de la mécanique analytique,* pt. 1, chap. 2. Rolle's first paper was presented at the Academy's sessions of 17 and 21 July 1700.

32. Pierre de Fermat, "Methodus ad disquirendam maximam et minimam" (c. 1637), in *Oeuvres de Fermat,* ed. Paul Tannery and Charles Henry (Paris: Gauthier-Villars, 1891–1912), 1:133–136. Fermat's algebraic method rested on the procedure of adequation *(adégalisation),* or equalization.

33. Hudde's letters and notes can be found in an edition of Descartes's *Geometry* published some twenty years after the original of 1637, *Geometria a Renato Des Cartes anno 1637 Gallice edita [. . .] (Johannis Huddenii epistolae duae,*

quarum altera de aequationem reductione, altera de maximis et minimis agit)
[. . .] (Amsterdam, 1659–1661), along with those of de Beaune, van Schooten,
and van Heuraet. Hudde's algebraic procedure involved determining double roots
for a polygon, which is to say that it gave not only the maxima and the minima
but also the points of intersection of the branches of the curves. In modern terms,
Hudde's method can be expressed with the help of the following two rules: first,
if b is a double root of f(x) = 0, then b is also a root of f'(x) = 0; and second,
if f(a) is the value of a maximum or of a minimum of a polynomial f(x), then
f'(a) = 0.

34. See Bernard Nieuwentijt, *Considerationes circa analyseos ad quantitates*
infinite parvas applicatae principia, et calculi differentialis usum in resolvendis
problematibus geometricis (Amsterdam, 1694); *Analysis infinitorum seu curvili-*
neorum proprietates ex polygonorum deductae (Amsterdam, 1695); *Consider-*
ationes secundae circa calculi differentialis principia, et responsio ad virum nobil-
issimum G. G. Leibnitium (Amsterdam, 1696).

35. Guillaume de l'Hospital, *Analyse des infiniment petits,* 2; the translation
is from *The method of fluxions both direct and inverse,* trans. E. Stone (London,
1730), 3.

36. This work, the full title of which is *The Analyst; or, a discourse addressed*
to an infidel mathematician. Wherein it is examined whether the object, prin-
ciples, and inferences of modern analysis are more distinctly conceived, or more
evidently deduced, than religious mysteries and points of faith, continued the bat-
tle waged in the *Alciphron; or, The minute philosopher* (London, 1732), against
the "minute philosophers" while extending it to a new group of free thinkers, the
"infidel mathematicians."

37. George Berkeley, *The Analyst* (London, 1734), in *The Works of George*
Berkeley, Bishop of Cloyne, ed. A. A. Luce and T. E. Jessop (London: Thomas
Nelson and Sons, 1948–1957), 4:67.

38. Ibid., 68.

39. Ibid., 69.

40. Ibid., 76.

41. Ibid., 74.

42. A. Ac. Sc. Registres, vol. 20, fols. 85r and 95r. Rolle's paper, "Troisième
remarque sur les principes de la géométrie des inf. petits," takes up folios 95r–
101r.

43. Ibid., fol. 95r.

44. Ibid., fol. 95v.

45. Ibid., fols. 95v–96r.

46. Ibid., fol. 96r.

47. Ibid., fol. 96v.

48. Ibid., fols. 235r–240v. This is Varignon's third written response to Rolle's
polemic: see Blay, "Deux moments de la critique du calcul infinitésimal."

49. Montucla, *Histoire des mathématiques,* 3:112.

50. A. Ac. Sc. Registres, vol. 20, fol. 237r–237v; and in Bernoulli, *Briefwech-*
sel, 2:271–281.

51. A. Ac. Sc. Registres, vol. 20, fol. 237v.

52. Compare the modern expression of Hudde's method given above in n. 33.

53. Bernoulli, *Briefwechsel,* 2:273.

54. See Blay, *La naissance de la mécanique analytique.*

55. Bernoulli, *Briefwechsel,* 2:29.

56. Pierre Varignon, "Démonstration générale de l'arithmétique des infinis ou de la géométrie des indivisibles," Archives of the Royal Academy of Sciences, Pochettes de Séances of 2 January 1694, 4 pp.

57. Ibid.

58. Ibid. This demonstration is supplemented by three corollaries.

59. Varignon, "Rectification et quadrature de l'évolute du cercle décrite à la manière de Monsieur Hugens," A. Ac. Sc. Registres, vol. 14, fol. 135r–135v. The three supplementary papers deposited the previous year in the Pochettes de Séances are, in order of presentation: "Réfutation du sentiment du P. Guldin et de MM. Wallis et Sturmius sur la longueur de la spirale d'Archimède" (13 March 1694), 2 pp. (the delivery of this paper was announced in A. Ac. Sc. Registres, vol. 14, fol. 10v); "Démonstration de six manières différentes de trouver les rayons des développées, lors même que les ordonnées des courbes qu'elles engendrent, concourent en quelque point que ce soit et par conséquent aussi pour le cas où elles sont parallèles" (27 November 1694), 3 pp. (this was a brief adaptation by Varignon of the paper by Jacques Bernoulli published in *AE* in June 1694); "Manière générale de trouver les tangentes spirales de tous les genres et de tant de révolutions qu'on voudra avec leurs quadratures indéfinies" (11 December 1694), 5 pp.

60. Varignon, "Courbe isochrone: Le long de laquelle descendent d'une vitesse uniforme par rapport à l'horizon, en sorte qu'ils s'en approchent également en temps égaux," A. Ac. Sc. Registres, vol. 14, fols. 135v–136r. Later in the same volume, at fol. 158r, it is indicated that Varignon delivered this paper on 30 July 1695.

61. Varignon, "Rectification et quadrature indéfinies des cycloïdes à bases circulaires, quelque distance qu'on suppose entre leur point décrivant et le centre de leur cercle mobile," ibid., fols. 181r–184r.

62. Varignon, "Du déroulement des spirales de tous les genres, où l'on fait voir qu'elles se déroulent toutes en paraboles d'un degré seulement plus haut que leur, avec une méthode générale pour toutes ces sortes de déroulement," ibid., fols. 192r–193r.

63. Varigon, "Regle generale pour toutes sortes de mouvemens de vitesses quelconques variées a discretion" and "Application de la Regle generale des vitesses variées, comme on voudra, aux mouvemens par toutes sortes de courbes, tant mecaniques que geometriques. D'ou l'on deduit encore une nouvelle maniere de démontrer les chutes isochrones dans la cycloïde renversée." These papers may be found in A. Ac. Sc. Registres, vol. 17, fols. 297v–305r and fols. 387r–391v, respectively; and in my article "Quatre mémoires inédits de Pierre Varignon consacrés à la science du mouvement," *Archives internationales d'histoire des sciences* 123 (1989): 218–248.

64. A. Ac. Sc. Registres, vol. 17, fol. 298r.

65. *AH,* year 1700 (1702): 85.

66. A. Ac. Sc. Registres, vol. 17, fol. 298r. The accompanying illustration is from the same source.

67. It will be recalled that in modern terms, using the concept of a function (x representing space; t, time; and v, speed), motion is defined by the time equation x = f(t), which yields the trajectory of a point x over time. The graph of the function x = f(t) is called the position diagram. The graph of the function v = f′(t) is called the velocity diagram. Varignon also introduces the function v = g(x) (expressed, again, in modern terms), which constitutes an alternative diagram of velocity.

68. A. Ac. Sc. Registres, vol. 17, fol. 298r. The idea of function is expressed here and elsewhere in Varignon's writing by the phrase *"quelle qu'elle soit"* (literally, "whatever [curve] it may be").

69. Varignon does not explicitly define the concept of an instant here. In a paper of July 1707 (see n. 72, below) he was more precise: "Definition I. By the word instant, we understand here an infinitely small particle of time, or (to speak as some modern [authors have done] since Descartes), indefinitely small, that is to say, less than whatever magnitude of time may be assigned: this is what in [the] language of the ancients was called *minor quavis quantitate data* [. . .]" (*AM,* year 1707 [1708]: 222).

70. A. Ac. Sc. Registres, vol. 17, fol. 298v.

71. Ibid., fols. 298v–299r.

72. Varignon, "Des mouvemens variées à volonté, comparés entr'eux et avec les uniformes," *AM,* year 1707 (1708): 223.

73. Ibid., 224.

74. In this connection see in particular the quite famous passage on the theory of measurement in Laplace's *Exposition du Système du Monde,* in *Oeuvres complètes de Laplace,* 6th ed. (Paris, 1835), vol. 6, bk. 3, chap. 2, p. 156.

75. Varignon, *Traité du mouvement et de la mesure,* 22.

76. A. Ac. Sc. Registres, vol. 17, fol. 299r.

77. Ibid., fol. 299r–299v.

78. Ibid., fol. 387r.

79. See de l'Hospital, *Analyse des infiniment petits,* 3.

80. KG represents, therefore, the distance traveled by the body along its trajectory GG. The point K, the origin of the motion along this trajectory, proves very useful later in the paper.

81. A. Ac. Sc. Registres, vol. 17, fol. 387v. In fact, this formula had already figured in the correspondence between Varignon and Jean Bernoulli in connection with the treatment of a particular case, that of a speed calculated according to Galileo's law of falling bodies, involved in the brachistochrone problem: see Bernoulli, *Briefwechsel,* vol. 2, letters 23 and 24.

82. Varignon had already worked on the problem of isochronic falls in the inverted cycloid at the Royal Academy during 1697–1698, but his approach remained very geometrical in spirit and did not rely on the concept of speed in each instant.

83. A. Ac. Sc. Registres, vol. 17, fol. 387v. The illustration accompanying this passage is taken from the same source.

84. Ibid., fol. 391r–391v.

85. See Pierre Varignon, "Des mouvemens faits dans des milieux qui leur résistent en raison quelconque," *AM,* year 1707 (1708): 382–476. For a detailed study

of these questions, see Michel Blay, "Varignon ou la théorie du mouvement des projectiles 'comprise en une proposition générale,'" *Annals of Science* 45, no. 6 (1988): 591–618, as well as *La naissance de la mécanique analytique*, 251–330.

86. *AM*, year 1707 (1708): 386.

87. Ibid. It needs to be kept in mind that Varignon never used what we refer to (in modern notation) as the definite integral of f(x)dx taken between the bounds a and b.

88. Varignon, "Manière générale de déterminer les forces, les vitesses, les espaces, et les temps, une seule de ces quatre choses étant données dans toutes sortes de mouvemens rectilignes variés à discrétion," *AM*, year 1700 (1703): 22–27; and A. Ac. Sc. Registres, vol. 19, fols. 31r–37r. There are some discrepancies between the two versions.

89. Varignon, "Du mouvement en générale par toutes sortes de courbes; et des forces centrales, tant centrifuges que centripètes, nécessaires aux corps qui les décrivent," *AM*, year 1700 (1703): 83–101; and A. Ac. Sc. Registres, vol. 19, fols. 133v–141v.

90. *AM*, year 1700 (1703): 23.

91. Ibid., 86. This expression can be interpreted directly in modern terms, given that mass does not enter into Varignon's calculations, as the statement in differential form of the theorem of kinetic energy.

92. Auguste Comte, *Philosophie première: Cours de philosophie positive, leçons 1 à 45* (Paris: Hermann, 1975), lesson 17, 268.

Epilogue

1. Although Fontenelle's nomination is not mentioned in A. Ac. Sc. Registres, beginning with the session of 17 November 1697 Fontenelle's signature replaces that of Du Hamel next to Abbé Bignon's signature at the end of the session minutes: see ibid., vol. 17, fol. 3r.

2. Fontenelle, having petitioned for emeritus status in 1730 and in 1737, was finally granted it in 1740. On this point see ibid., vol. 59, fols. 213, 215, 217, 228. His duties were taken over by de Mairan at the Saturday session of 7 January 1741 (ibid., vol. 60, fol. 1r).

3. The first part, pp. 1–392, divided into twelve sections, bears the title "General System of the Infinite" (Systeme générale de l'infini); the second, pp. 393–546, divided into eight sections, bears the title "Various Applications and Remarks" (Différentes applications ou remarques). The three final pages of the work (pp. 546–548) are entitled "Comment on the Sums of Series" (Réflexion sur les sommes des suites). This "comment" appears to have been added after the work had been delivered to the printers; work on sums of series had been presented to the Royal Academy by Nicole at its session of 25 June 1727 (see A. Ac. Sc. Registres, vol. 46, fols. 239, 240–246). See also my two articles on Fontenelle's *Elements:* "Du fondement du calcul différentiel au fondement de la science du mouvement dans les 'Élémens de la géométrie de l'infini' de Fontenelle," *Studia Leibnitiana* 17 (1989): 99–122; and "Du système de l'infini au statut des nombres incommensurables dans les 'Élémens de la géométrie de l'infini' de Fontenelle," in *Le labyrinthe du continu: Colloque de Cérisy*, ed. Jean-Michel Salanskis and Hourya Sina-

ceur (Paris: Springer-Verlag, 1992), 61–75. See also the second part of chapter 3 of my *La naissance de la mécanique analytique;* and, more generally, the new edition of Fontenelle's *Elements* that I have recently edited with Alain Niderst (Paris: Klincksieck, 1995).

4. At the Wednesday session of 14 August 1726, a commission composed of Dortous de Mairan and François Nicole was charged by the Royal Academy, at Fontenelle's request, with responsibility "for examining my Elements of the Geometry of the Infinite" (A. Ac. Sc. Registres, vol. 45, fol. 255r). A summary of the findings of this commission was presented to the Academy at its Saturday session of 22 February 1727 (ibid., vol. 46, fols. 73–74). The full report drafted by the commission was published by Abbé Trublet, to whom de Mairan had communicated it, in his *Mémoires pour servir à l'histoire de la vie et des ouvrages de Fontenelle tirés du Mercure de France, 1756, 1757, et 1758, par l'abbé Trublet, seconde édition corrigée et augmentée* (Amsterdam, 1759).

5. See Bernard Le Bovier de Fontenelle, *Élemens de la géométrie de l'infini* (Paris, 1727), 18–19.

6. Ibid., 9–10.

7. In this connection the significance of the word "Elements" in the title of Fontenelle's book should be stressed.

8. Ibid., 15–16.

9. Ibid., 16. "Paradox," as used in this passage, should not be understood in too technical a sense, but rather in a way more in keeping with its etymology, as something that runs counter to common sense.

10. This is the title given by Fontenelle to the first part of his book; see n. 3, above.

11. Fontenelle, *Élemens,* 18–19.

12. Ibid., 19–20.

13. Leibniz, GM, 4:110. See also the undated reply from Leibniz to Fontenelle's letter of 9 September 1704, in L.-A. Foucher de Careil, ed., *Lettres et opuscules inédits de Leibniz* (Paris, 1854), 234.

14. Bernard de Fontenelle, *Oeuvres de M. de Fontenelle* (Paris, 1766), 11:157. Shortly afterward, Père Castel published an excerpt of the *Elements* in the *Journal de Trévoux.*

15. See the letters exchanged between Fontenelle and Bernoulli held in the archives of the Öffentliche Bibliothek der Universität, Basel (hereafter UB Basel), MS LIa 692. On this correspondence see Michel Blay, "Note sur la correspondence entre Jean I Bernoulli et Fontenelle," *Corpus* 13 (1989): 93–100.

16. UB Basel, MS LIa 692.

17. See in particular Fontenelle's letters dated 7 June 1725, 8 May 1729, 28 June 1729, and 29 August 1729, UB Basel, MS LIa 692.

18. See the letters dated 20 November 1728 and 29 March 1729, published by Jacqueline de la Harpe in her article "Des inédits de Fontenelle: Sa correspondence avec J.-P. Crousaz," *Revue historique vaudoise* (June 1954): 90–108.

19. See the letter dated 7 April 1730 in *Oeuvres de M. de Fontenelle,* 11:40–41.

20. See the letter dated 22 September 1739, in ibid., 11:29–30.

21. On this point see also Fontenelle's "Eulogy of Jacques Bernoulli," *AH,* year 1705 (1706): 139–150.

22. Fontenelle, *Élemens,* 11.

23. Ibid., 13.

24. Ibid., 14.

25. See n. 32, below.

26. Fontenelle, *Élemens,* 1.

27. Ibid., 29.

28. Section II, "De la grandeur infiniment grande," ibid., 29–57.

29. Brunschvicg, *Les étapes de la philosophie mathématique,* 244.

30. Fontenelle, *Élemens,* 11.

31. Ibid., 30–31.

32. Ibid., 30. The symbol ∞ was borrowed by Fontenelle from Wallis, who had used it previously in part I, proposition 1, of his *De sectionibus conicis* (London, 1655). Fontenelle's infinite number, which finally (and this is one of its weaknesses) is part of the "natural Series," is therefore not to be confused with Cantor's number ω in the *Grundlagen einer allgemeinen Mannichfaltigkeitslehre* (Leipzig, 1883). On the other hand, certain points of contact can be demonstrated between Fontenelle's treatment and more recent developments connected with Robinson's nonstandard analysis. In this connection see the appendix of H. J. M. Bos's 1974 article "Differentials, Higher-Order Differentials and the Derivative in the Leibnizian Calculus," 81–86; Abraham Robinson, *Non-standard Analysis* (Amsterdam: North-Holland, 1966); and Edward Nelson, "Internal Set Theory: A New Approach to NSA," *Bulletin of the American Mathematical Society* 83 (1977): 1165–1198.

33. Fontenelle, *Élemens,* 30–31.

34. Ibid., 31.

35. Ibid., 35. Note that in the preceding paragraphs Fontenelle has made it quite clear, with the help of examples, what is to be understood by "orders of infinites."

36. See the first day of Galileo's *Discorsi,* as well as the commentary in Louis Couturat, *De l'infini mathématique* (Paris: Blanchard, 1973), 445–446.

37. Fontenelle, *Élemens,* 59.

38. Ibid.

39. Ibid., 61.

40. Ibid., 63.

41. Ibid., 63–64. In his letter to Jean Bernoulli dated 22 April 1725, Fontenelle argues on the basis of infinitely small quantities; see UB Basel, MS LIa 692, and my previously cited article on their correspondence.

42. Fontenelle, *Élemens,* 64.

43. Fontenelle himself uses the term "paradox" to characterize his conclusion (ibid.).

44. Fontenelle actually gives seven reasons, which for the most part are summarized by the two given here.

45. See ibid., 30–31.

46. Ibid., 65.

47. Ibid., 66.

48. Ibid., 66–67.

49. Ibid., 74, 82.

50. UB Basel, MS LIa 676, fols. 114–118.

51. Brunschvicg, *Les étapes de la philosophie mathématique*, 244.

52. Note that section IV of the first part of the *Elements* (pp. 116–146) is entitled "Of Infinitely Small Magnitude."

53. Ibid., 146.

54. "Sur les suites infinies de grandeurs quelconques," ibid., 184–244.

55. Ibid., 243.

56. Ibid.

57. Ibid.

58. "Des variations et des changements des courbes," ibid., 271–310.

59. Ibid., 310.

60. "Règles générales pour déterminer par le calcul différentiel tout ce qui appartient au cours d'une courbe rapportée à un axe," ibid., 311–352.

61. Section XII is entitled "General Rule for Determining the Curvature of Curves by the Differential Calculus" (Règle générale pour déterminer par le calcul différentiel, la courbure des courbes), ibid., 353–392. The second part of the *Elements* follows, in eight sections, under the title "Various Applications and Remarks" (Différentes applications ou remarques).

62. See the eulogy of Varignon in *AH,* year 1722 (1724): 141.

63. Ibid., 142.

64. *AH,* year 1699 (1702): 100.

65. See the discussion in chap. 1, above.

66. See the discussion in chap. 4, above.

67. *AH,* year 1701 (1704): 80–83.

68. *AH,* year 1703 (1705): 76.

69. Ibid., 65.

70. *AH,* year 1707 (1708): 58.

71. See "Sur les forces des corps en général," in Fontenelle, *Élemens,* 516–546.

72. Ibid., 524–525.

73. Ibid., 525–526.

74. Edmund Husserl, *La crise des sciences européennes et la phénoménologie transcendentale,* trans. Gérard Granel (Paris: Gallimard, 1976), 54.

75. See the foreword to the 1788 edition of Lagrange, *Mécanique analitique.*

76. Ernst Mach, *The Science of Mechanics: A Critical and Historical Account of Its Development,* trans. Thomas J. McCormack, 6th ed. (with revisions through the 9th German ed.) (Lasalle, Ill.: Open Court, 1974), 561.

77. Lagrange, *Mécanique analitique,* 1.

78. Ibid., 207.

79. "Formule générale de la statique pour l'équilibre d'un système quelconque de forces, avec la manière de faire usage de cette formule," ibid., 24ff.

80. Ibid., 24. We speak today of the principle of potential energies, stating it as follows: "The potential energy of all forces, including those of inertia, applied

to a frictionless mechanical system which they maintain in a state of equilibrium, is zero for every potential displacement."

81. Ibid.

82. Ibid., 25.

83. Ibid., 26.

84. Ibid.

85. Ibid., 233. Note that in this second section of the second part of Lagrange's work ("Formule générale de la dynamique pour le mouvement d'un système de corps animés par des forces quelconques"), expressions such as the last two do not imply that the forces P, Q, R, etc. are central forces. The centers in question here are any points taken along the directional axes of the forces.

86. Ibid., 234.

87. Ibid.

88. The phrase is Husserl's; see n. 74, above.

89. Mach, *Science of Mechanics,* 577.

90. Ibid., 561–562.

91. Hoené Wronski, *La philosophie de l'infini: Contenant des contre-refléxions sur la métaphysique du calcul infinitesimal* (Paris, 1814). Shortly after his arrival in France, Wronski published his first important work, *Philosophie critique, fondée sur le premier principe du savoir humain* (Marseilles, 1803). For further detail on the extremely rich mathematical and philosophical work of this author, see the entry in the *Dictionary of Scientific Biography,* ed. Charles C. Gillispie (New York: Scribners, 1970–1976), vol. 15 (suppl. 1), 225–226.

92. Lazare Carnot presented his method of eliminating errors in his *Réflexions sur la métaphysique du calcul infinitesimal* (1797) (English ed.: *Reflexions on the Metaphysical Principles of the Infinitesimal Analysis,* trans. W. R. Browell [Oxford, 1832]). Lagrange attempted to found analysis neither on the method of limits nor on that of infinitely small quantities, but on algebraic methods, relying in particular on the series developed by Brook Taylor (1685–1731), in his *Théorie des fonctions analytiques* (1797). A survey of these debates may be found in A. Guerraggio and M. Panza, "Le reflexions di Carnot e le contre-reflexions di Wronski," *Epistemologia* 8 (1985): 3–22.

93. Wronski, *Philosophie de l'infini,* 34.

94. Ibid., 35.

95. Ibid., 36.

96. Ibid., 35–36.

97. Joseph-Louis Lagrange, *Théorie des fonctions analytiques, contenant les principes du calcul différentiel, dégagés de toute considération d'infiniment petits, d'évanouissants, de limites et de fluxions, et réduits à l'analyse algébrique des quantités finies* (Paris, 1797).

98. Carnot, *Reflexions,* 39–41.

99. The reader will recall the passage in the Marquis de l'Hospital's *Analyse des infiniment petits* cited earlier (see chap. 4, n. 35, above), which Varignon also drew upon in constructing the concept of speed in each instant: "Grant that two Quantities, whose Difference is an infinitely small Quantity may be taken (or used) indifferently for each other: or (which is the same thing) that a Quantity,

which is increased or decreas'd only by an infinitely small Quantity, may be considered'd as remaining the same."

100. Wronski, *Philosophie de l'infini*, 37.

101. Ibid., 38.

102. Ibid., 38–39.

103. Ibid., 40–41.

104. Ibid., 43. Hippolyte Margerin (1799–?), a follower of Saint-Simon, adopted Wronski's thesis and expanded upon it: "The infinite is not only the most powerful instrument of mathematical investigation, it is the very reason for the existence of science" (Bibliothèque de l'Arsenal, Fonds Enfantin, MS 7644, fol. 336r). I am grateful to Philippe Régnier for bringing this passage to my attention.

Académie Royale des Sciences (Paris). Archives. Registres manuscrits des procès-verbaux des séances de l'Académie Royale des Sciences de Paris.

———. *Histoire de l'Académie Royale des Sciences.* Vol. 1, *Depuis son établissement en 1666 jusqu'à 1686.* Vol. 2, *Depuis 1686 jusqu'à son renouvellement en 1699.* Paris, 1733.

———. *Histoire de l'Académie Royale des Sciences avec les Mémoires de Mathématique et de Physique pour la même année: Tirés des Registres de cette Académie.* 92 vols. Paris, 1702–1797.

———. *Mémoires de l'Académie Royale des Sciences depuis 1666 jusqu'à 1699.* 9 vols., numbered 3–11 (following vols. 1–2 of the *Histoire;* see above). Paris, 1730.

Aiton, Eric J. *The Vortex Theory of Planetary Motions.* New York: Elsevier, 1972.

Andersen, K. "Cavalieri's Method of Indivisibles." *Archive for History of Exact Sciences* 31 (1985): 291–368.

Archimedes. *On Floating Bodies.* In *Works,* 253–300.

———. *On the Equilibrium of Planes.* In *Works,* 189–220.

———. *Opera Omnia.* Edited by J. L. Heiberg. 2d ed. 3 vols. Leipzig: Teubner, 1910–1915.

———. *The Works of Archimedes.* Edited by T. L. Heath. Cambridge: Cambridge University Press, 1897. Reprint, New York: Dover Publications, 1912, 1950.

———. "The Method of Indivisibles: Changing Understanding." *Studia Leibnitiana,* Special Issue 14 (1986): 14–25.

Aristotle. *The Complete Works of Aristotle.* Revised Oxford translation. Edited by Jonathan Barnes. 2 vols. Bollingen Series, no. 71. Princeton: Princeton University Press, 1984.

Bachelard, Gaston. *L'activité rationaliste de la physique contemporaine.* Paris: Presses Universitaires de France, 1951.

Bachelard, Suzanne. *La conscience de rationalité: Étude phénoménologique sur la physique mathématique.* Paris: Presses Universitaires de France, 1958.

Baron, Margaret E. *The Origins of the Infinitesimal Calculus.* Oxford: Pergamon, 1969. Reprint, New York: Dover Publications, 1987.

Barrow, Isaac. *Lectiones geometricae: in quibus (praesertim) generalia curvarum linearum symptomata declarantur.* Cambridge, 1670.

Barthélemy, Georges. "Concepts et méthodes de la mécanique rationelle dans les *Principia* de Newton." Doctoral dissertation, Sorbonne, Paris, 1985.

Beeckman, Isaac. *Journal, 1604–1634.* Edited by Cornelis de Waard. 4 vols. The Hague: M. Nijhoff, 1939–1953.

Bell, A. E. *Christiaan Huygens and the Development of Science in the Seventeenth Century.* New York: Longmans, Green, 1947.

Berkeley, George. *Alciphron; or, The minute philosopher.* London, 1732.

———. *The Analyst; or, a discourse addressed to an infidel mathematician. Wherein it is examined whether the object, principles, and inferences of modern analysis are more distinctly conceived, or more evidently deduced, than religious mysteries and points of faith.* London, 1734.

———. *Oeuvres.* Edited by Geneviève Brykman. 3 vols. Paris: Presses Universitaires de France, 1985–1992.

———. *The Works of George Berkeley, Bishop of Cloyne.* Edited by A. A. Luce and T. E. Jessop. 9 vols. London: Thomas Nelson and Sons, 1948–1957.

Bernhardt, Jean. "La constitution de la Théorie de la Science chez Thomas Hobbes." 2 vols. Doctoral thesis, Sorbonne, Paris, 1983.

———. "Infini, substance et attributs: Sur le spinozisme." *Cahiers Spinoza* 2 (1978): 53–92.

Bernier, François. *Abrégé de la philosophie de Gassendi.* 8 vols. Lyons, 1678. 2d ed., 7 vols., 1684.

Bernoulli, Daniel. *Die Werke von Daniel Bernoulli.* Edited by David Spieser et al. 8 vols. planned. Basel: Birkhäuser, 1982–.

———. *Die Werke von Daniel Bernoulli.* Vol. 3 (Mechanics). Edited by D. Spieser, A. de Baenst-Vandenbroucke, J.-L. Pietenpol, and P. Radelet de Grave. Basel: Birkhäuser, 1987.

Bernoulli, Jacques. "Analysis problematis antehac propositi: De Inventione Lineae descensus a corpore gravi percurrendae uniformiter, sic ut temporibus aequalibus aequales altitudines emetiatur; et alterius cujusdam problematis Proposito." *Acta Eruditorum* (May 1690): 217–220.

———. "Curvae dia-causticae." *Acta Eruditorum* (June 1693): 244–256.

———. "Curvatura laminae elasticae." *Acta Eruditorum* (June 1694): 262–276.

———. "Curvatura veli." *Acta Eruditorum* (May 1692): 202–211.

———. *Jacobi Bernoulli, Basileensis, Opera.* 2 vols. Geneva, 1744. Reprint, Brussels: Culture et Civilisation, 1967.

Bernoulli, Jean. *Der Briefwechsel von Johann I Bernoulli.* Vol. 1 (Correspondence with the Marquis de l'Hospital) edited by O. Spiess. Vols. 2/1–2 (Correspondence with Varignon) edited by P. Costabel and J. Peiffer. Basel: Birkhäuser Verlag, 1955–1992.

———. *G. Leibnitii et Johann Bernoulli commercium philosophicum et mathematicum.* 2 vols. Geneva, 1745.

———. *Johannis Bernoulli [. . .], Opera Omnia tam antea sparsim edita, quam hactenus inedita.* 4 vols. Lausanne and Geneva, 1742. Reprint, Hildesheim: Georg Olms, 1968.

Bernstein, H. "Conatus, Hobbes, and the Young Leibniz." *Studies in History and Philosophy of Science* 11 (1980): 25–37.

Bertoloni Meli, Domenico. "Leibniz's Excerpts from the *Principia Mathematica*." *Annals of Science* 45 (1988): 477–505.

———. "The Relativization of Centrifugal Force." *Isis* 81 (1990): 21–43.

Blay, Michel. "Deux moments de la critique du calcul infinitésimal: Michel Rolle et George Berkeley." *Revue d'histoire des sciences* 39 (1986): 223–253.

———. "Du fondement du calcul différentiel au fondement de la science du mouvement dans les 'Élémens de la géométrie de l'infini' de Fontenelle." *Studia Leibnitiana* 17 (1989): 99–122.

———. "Du système de l'infini au statut des nombres incommensurables dans les 'Élémens de la géométrie de l'infini' de Fontenelle." In *Le labyrinthe du continu: Colloque de Cérisy*, edited by Jean-Michel Salanskis and Hourya Sinaceur, 61–75. Paris: Springer-Verlag, 1992.

———. *La naissance de la mécanique analytique*. Paris: Presses Universitaires de France, 1992.

———. "Note sur la correspondence entre Jean I Bernoulli et Fontenelle." *Corpus* 13 (1989): 93–100.

———. *Les "Principia" de Newton*. Paris: Presses Universitaires de France, 1995.

———. "Quatre mémoires inédits de Pierre Varignon consacrés à la science du mouvement." *Archives internationales d'histoire des sciences* 123 (1989): 218–248.

———. "Recherches sur les forces exercées par les fluides en mouvement à l'Académie Royale des Sciences: 1668–1669." In *Mariotte, savant et philosophe: Analyse d'une renommée*, edited by Pierre Costabel and Michel Blay, 91–124. Paris: Vrin, 1986.

———. "Le traitement newtonien du mouvement des projectiles dans les milieux résistants." *Revue d'histoire des sciences* 40, nos. 3–4 (1987): 325–355.

———. "Varignon et le statut de la loi de Torricelli." *Archives internationales d'histoire des sciences* 35, nos. 114–115 (1985): 330–345.

———. "Varignon ou la théorie du mouvement des projectiles 'comprise en une proposition générale.'" *Annals of Science* 45, no. 6 (1988): 591–618.

Blay, Michel, and Georges Barthélemy. "Changements de repères chez Newton: Le problème des deux corps dans les *Principia*." *Archives internationales d'histoire des sciences* 34, no. 112 (1984): 69–98.

Bortolotti, Ettore. "L'oeuvre géométrique d'Évangeliste Torricelli." Translated by P. Souffrin and J.-P. Weiss. In De Gandt, ed., *L'oeuvre de Torricelli*, 115–146.

Bos, H. J. M. "Differentials, Higher-Order Differentials and the Derivative in the Leibnizian Calculus." *Archive for History of Exact Sciences* 14, no. 1 (1974): 1–90.

———. "Fundamental Concepts of the Leibnizian Calculus." *Studia Leibnitiana*, Special Issue 14 (1986): 103–118.

Boyer, Carl B. *The History of the Calculus and Its Historical Development*. New York: Hafner, 1949. Reprint, New York: Dover Publications, 1959.

Brandt, Frithiof. *Thomas Hobbes' Mechanical Conception of Nature*. Translated by Vaughan Maxwell and Annie I. Fausboll. Copenhagen: Levin and Munksgaard, 1921. 2d ed., Copenhagen: Levin and Munksgaard, 1928.

Brunet, Pierre. *L'introduction des Théories de Newton en France au XVIIIᵉ siècle*. Paris: Blanchard, 1931. Reprint, Geneva: Slatkine, 1970.

Bruno, Giordano. *De l'infinito, universo e mondi*. London, 1584.

Brunschvicg, Léon. *Les étapes de la philosophie mathématique*. Paris: Alcan, 1912. Reprint, Paris: Blanchard, 1972.

———. *L'expérience humaine et la causalité physique*. Paris: Presses Universitaires de France, 1949.

Bucciantini, Massimo, and Maurizio Torrini, eds. *Geometria et atomismo nella scuola galileinana*. Florence: Olschki, 1992.

Burtt, Edwin A. *The Metaphysical Foundations of Modern Science*. New York: Harcourt, Brace, 1925. 4th ed., New York: Anchor Books, 1954.

Cajori, Florian. *A History of Mathematical Notations*. 2 vols. Chicago: Open Court, 1929. 2d ed., 1930. 3d ed., 1952.

———. *A History of the Conceptions of Limits and Fluctuations in Great Britain from Newton to Woodhouse*. Chicago: Open Court, 1919.

Cantor, Georg. *Grundlagen einer allgemeinen Mannichfaltigkeitslehre*. Leipzig, 1883.

Carnot, Lazare. *Reflexions on the Metaphysical Principles of the Infinitesimal Analysis*. Translated by W. R. Browell. Oxford, 1832.

Carré, Jean-Raoul. *La philosophie de Fontenelle ou le sourire de la raison*. Paris: Alcan, 1932. Reprint, Geneva: Slatkine, 1970.

Carré, Louis. *Méthode pour la mesure des surfaces; la dimension des solides, leurs centres de pesanteur, de percussion et d'oscillation, par l'application du calcul intégral*. Paris, 1700.

Cassirer, Ernst. *Substanzbegriff und Funktionsbegriff: Untersuchungen über die Grundfragen der Erkenntniskritik*. Berlin: B. Cassirer, 1910.

Catelan, Abbé de. *Logistique pour la Science générale des lignes courbes, ou Manière universelle et infinie d'exprimer et de comparer les puissances des grandeurs*. Paris, 1691.

———. *Principe de la science générale des lignes courbes, ou des principaux élémens de la géométrie universelle*. Paris, 1691.

Cavalieri, Bonaventura. *Carteggio* [Correspondence]. Edited by Giovanna Baroncelli. Florence: Olschki, 1987.

———. *Exercitationes geometricae sex*. Bologna, 1647. Reprint, Rome: Cremonese, 1980.

———. *Geometria indivisibilibus continuorum nova quadam ratione promota*. Bologna, 1635.

———. "Opere inedite." Edited by Sandra Giuntini, Enrico Giusti, and Elisabetta Ulivi. *Bolletino di storia della scienze matematiche* 5 (1985): 1–352.

———. *Lo specchio ustorio, overo trattato delle settioni conichi*. Bologna, 1632.

Caveing, Maurice. *Zénon d'Élée, Prolégomènes aux doctrines du continu: Étude historique et critique des fragments et témoignages*. Paris: Vrin, 1982.

Child, J. M. *The Early Mathematical Manuscripts of Leibniz, Translated from the Latin Texts*. Chicago: Open Court, 1920.

Clagett, Marshall. *The Science of Mechanics in the Middle Ages*. Madison: University of Wisconsin Press, 1959.

Clavelin, Maurice. "Conceptual and Technical Aspects of the Galilean Geometrization of the Motion of Heavy Bodies." In Shea, ed., *Nature Mathematized*, 23–50.

———. *The Natural Philosophy of Galileo*. Translated by A. J. Pomerans. Cambridge: MIT Press, 1974.

———. *La philosophie naturelle de Galilée*. Paris: Colin, 1968. Reprint, Paris: Albin Michel, 1996.

Cohen, I. Bernard. *Introduction to Newton's "Principia."* Cambridge: Cambridge University Press; Cambridge, Mass.: Harvard University Press, 1971.

Comte, Auguste. *Philosophie première: Cours de philosophie positive, leçons 1 à 45*. Paris: Hermann, 1975.

———. *Physique sociale: Cours de philosophie positive, leçons 46 à 60*. Paris: Hermann, 1975.

Copernicus, Nicholas. *De revolutionibus orbium coelestium*. Translated by John F. Dobson and Selig Brodetsky. London: Royal Astronomical Society, 1947.

———. *Nicholas Copernicus: On the Revolutions*. Edited by Jerzy Dobrzycki and translated by Edward Rosen. Baltimore: Johns Hopkins University Press, 1978.

Costabel, Pierre. *Démarches originales de Descartes savant*. Paris: Vrin, 1982.

———. *Leibniz and Dynamics: The Texts of 1692*. Translated by R. E. W. Maddison. Ithaca: Cornell University Press, 1973.

———. "Liste des publications de Pierre Costabel." *Revue d'histoire des sciences* 43, no. 1 (1990): 313–324.

———. "Pierre Varignon (1654–1722) et la diffusion en France du calcul différentiel et intégral." *Conférences du Palais de la Découverte*, series D, no. 108 (4 December 1965): 1–28.

———. *La question des forces vives*. Cahiers d'histoire et de philosophie des sciences, n. s., no. 8.. Paris: Centre National de la Recherche Scientifique, Centre de Documentation Sciences Humaines, 1983.

Couturat, Louis. *De l'infini mathématique*. Paris, 1896. Reprint, Paris: Blanchard, 1973.

Craig, John. *Methodus figurarum lineis rectis et curvis comprehensarum quadraturas determinandi*. London, 1685.

———. *Methodus mathematicus de figurarum curvilinearum quadraturis et locis geometricis*. London, 1693.

Cudworth, Ralph. *The True Intellectual System of the Universe*. London, 1678.

De Gandt, François. *Force and Geometry in Newton's "Principia."* Translated by Curtis Wilson. Princeton: Princeton University Press, 1995.

———. "Force et géométrie: La théorie newtonienne de la force centripète présentée dans son contexte." 2 vols. Doctoral thesis, Sorbonne, Paris, 1987.

———. "Les indivisibles de Torricelli." In De Gandt, ed., *L'oeuvre de Torricelli*, 151–206.

———. "Le style mathématique des *Principia* de Newton." *Revue d'histoire des sciences* 39, no. 3 (1986): 195–222.

———, ed. *L'oeuvre de Torricelli: Science galiléenne et nouvelle géométrie*. Nice: Presses de l'Université de Nice, 1989.

Delorme, Suzanne, ed. *Galilée: Aspects de sa vie et de son oeuvre*. Paris: Presses Universitaires de France, 1968.

———. "La géométrie de l'Infini et ses commentateurs de Jean Bernoulli à M. de Cury." *Revue d'histoire des sciences* 10, no. 1 (1957): 339–359.

Desanti, Jean-Toussaint. *La philosophie silencieuse ou critique des philosophies de la science.* Paris: Seuil, 1975.

Desargues, Girard. "Brouillon project" (1639). In *L'oeuvre mathématique de G. Desargues,* edited by René Taton. Paris: Presses Universitaires de France, 1951. Reprint, Paris: Vrin, 1981.

Descartes, René. *Descartes, Principles of Philosophy.* Translated by V. R. Miller and R. P. Miller. Dordrecht: D. Reidel, 1983.

———. *Discours de la méthode: pour bien conduire sa raison, & chercher la vérité dans les sciences; Plus La dioptrique; Les météores; Et La géométrie. Qui sont des essais de cette méthode.* Leyden, 1637.

———. *Geometria a Renato Des Cartes anno 1637 Gallice edita [. . .] (Johannis Huddenii epistolae duae, quarum altera de aequationem reductione, altera de maximis et minimis agit) [. . .].* Amsterdam, 1659–1661.

———. *The "Geometry" of René Descartes.* Translated by David Eugene Smith and Marcia L. Latham. Chicago: Open Court, 1925. Reprint, New York: Dover Publications, 1954.

———. *Le Monde, ou Traité de la lumière* (Paris, 1664). Translated by Michael Sean Mahoney. New York: Abaris Books, 1979.

———. *Oeuvres.* Edited by Charles Adam and Paul Tannery. 12 vols., plus supplement. Paris: Cerf, 1896–1913. Reprint, Paris: Vrin/Centre National de la Recherche Scientifique, 1964–1974.

———. *The Philosophical Writings of Descartes.* Translated by John Cottingham, Robert Stoothof, and Dugald Murdoch. 3 vols. Cambridge: Cambridge University Press, 1984–1991.

———. *Principia Philosophiae.* Paris, 1644. In *Oeuvres,* vol. 9, pt. 2.

———. *Principles of Philosophy.* In *Philosophical Writings,* 1:177–292.

———. *Rules for the Direction of the Mind.* In *Philosophical Writings,* 1:7–78.

Drake, Stillman. *Galileo.* New York: Oxford University Press, 1980.

———. *Galileo at Work: His Scientific Biography.* Chicago: University of Chicago Press, 1978. Reprint, New York: Dover Publications, 1995.

———. *Galileo: Pioneer Scientist.* Toronto: University of Toronto Press, 1990.

Dugas, René. *Histoire de la mécanique.* Neuchâtel: Éditions du Griffon, 1950.

———. *La mécanique au XVIIᵉ siècle: Des antécédants scolastiques à la pensée classique.* Neuchâtel: Éditions du Griffon, 1954.

Du Hamel, Jean-Baptiste. *Regiae Scientiarum Academiae Historia.* Paris, 1698. 2d ed., Paris, 1701.

Dumont, Jean-Paul, Daniel Delattre, and Jean-Louis Poirier, eds. *Les Présocratiques.* Paris: Gallimard, 1988.

Dupont, P., with the assistance of S. Roero. "Leibniz 1684." *Quaderni di Matematica* 56 (December 1983): 1–99.

Earman, J. "Infinities, Infinitesimals, and Indivisibles: The Leibnizian Labyrinth." *Studia Leibnitiana* 7 (1975): 236–251.

Edwards, C. H., Jr. *The Historical Development of the Calculus.* New York: Springer-Verlag, 1979.

Euclid. *"L'optique" et "La cataoptrique."* Translated, with introduction and notes, by Paul Ver Eecke. Paris and Bruges: Desclée, De Brouwer, 1938. Reprint, Paris: Blanchard, 1959.

———. *The Thirteen Books of Euclid's "Elements."* Translated, with introduction and commentary, by Thomas L. Heath. 2d rev. ed. 3 vols. New York: Dover Publications, 1956.

Fermat, Pierre de. *Oeuvres de Fermat.* Edited by Paul Tannery and Charles Henry. 4 vols. Paris: Gauthier-Villars, 1891–1912.

Fleckenstein, J. O., P. Costabel, J. Peiffer, and B. Bilodeau, eds. "Liste des oeuvres de Pierre Varignon." In *Der Briefwechsel von Johann I Bernoulli,* 2:387–408.

Fontenelle, Bernard Le Bovier de. *Élemens de la géométrie de l'infini.* Paris, 1727.

———. *Elements de la géométrie de l'infini.* Edited and with an introduction by Michel Blay and Alain Niderst. Paris: Klincksieck, 1995.

———. *Oeuvres de M. de Fontenelle.* 11 vols. Paris, 1766.

Froidment, Libert. *Labyrinthus sive de compositione continui, liber unus.* Antwerp, 1631.

Gabbey, A. "Force and Inertia in Seventeenth Century Dynamics." *Studies in the History and Philosophy of Science* 2 (1971): 1–67.

Galilei, Galileo. *Dialogo [. . .] sopra i due Massimî Sistemi del Mondo, tolemaico e copernicano.* Florence, 1632. In *Opere,* vol. 7.

———. *Dialogue concerning the Two Chief World Systems—Ptolemaic and Copernican.* Translated by Stillman Drake. Berkeley and Los Angeles: University of California Press, 1953. Rev. ed., 1962.

———. *Discorsi e dimostrazione matematiche intorno a due nuove scienze.* Leyden, 1638. In *Opere,* 8:43–313.

———. *Galileo's Early Notebooks: The Physical Questions.* Translated from the Latin with historical and paleographical commentary by William A. Wallace. Notre Dame: University of Notre Dame Press, 1977.

———. *Les nouvelles pensées de Galilée traduit de l'italien en français par le P. Marin Mersenne.* Edited by Pierre Costabel and Michel-Pierre Lerner. Paris: Vrin, 1973.

———. *Opere.* Edizione nazionale. Edited by Antonio Favoro, Isidoro Del Lungo, and Valentino Cerruti. 21 vols. Florence: Barbera, 1890–1909.

———. *Le opere dei discepoli di Galileo Galilei: Carteggio, Volume Primo (1642–1649).* Edited by Paolo Galluzzi and Maurizio Torrini. Florence: Giunti-Barbera, 1975.

———. *Il Saggiatore.* Rome, 1623. In *The Controversy on the Comets of 1618,* translated by Stillman Drake and C. D. O'Malley, 183–184. Philadelphia: University of Pennsylvania Press, 1960.

———. *Two New Sciences.* Translated, with introduction and notes, by Stillman Drake. Madison: University of Wisconsin Press, 1974.

Galluzzi, Paolo. *Momento: Studi galileiani.* Rome: Edizione dell'Ateneo and Bizzarri, 1979.

Gassendi, Pierre. *Opera Omnia in sex tomos divisa.* 3 vols. Lyons, 1658.

Gillispie, Charles C., ed. *Dictionary of Scientific Biography.* 16 vols. New York: Scribners, 1970–1980.

Giusti, Enrico. "Aspetti matematici della cinematica galileiana," *Bollettino di storia della scienze matematiche* 5, no. 2 (1981): 3–42.

———. *Bonaventura Cavalieri and the Theory of Indivisibles.* Bologna: Cremonese, 1980.

Greenberg, J. L. "Mathematical Physics in Eighteenth Century France." *Isis* 77 (1986): 59–78.

Grégoire, Franz. *Fontenelle: Une philosophie désabusée.* Paris: Vrin, 1947.

Grégoire de Saint-Vincent. *Opus geometricum quadraturae circuli et sectioni coni.* Antwerp, 1647.

Guerraggio, A., and M. Panza. "Le reflexions di Carnot e le contre-reflexions di Wronski." *Epistemologia* 8 (1985): 3–22.

Hallyn, Fernand. *La structure poétique du monde: Copernic, Kepler.* Paris: Seuil, 1987.

Hannequin, Arthur. *Études d'histoire des sciences et d'histoire de la philosophie.* 2 vols. Paris: Alcan, 1908.

Helbing, Mario Otto. *La filosofia di Francesco Buonamici, professore di Galileo a Pisa.* Pisa: Nistri-Lischi, 1989.

Herivel, John. *The Background to Newton's "Principia."* Oxford: Clarendon Press, 1965.

———. "Newton's Achievement in Dynamics." In Palter, ed., *The "Annus Mirabilis" of Sir Isaac Newton,* 120–135.

Hess, Heinz-Jürgen. "Zur Vorgeschichte der 'Nova Methodus' (1676–1684)." *Studia Leibnitiana,* Special Issue 14 (1986): 64–102.

Hobbes, Thomas. *De Principiis et Ratio-cinatione geometrarum.* London, 1666.

———. *Elementorum philosophiae sectio prima De Corpore.* London, 1655.

———. *Elements of Philosophy, the first section: concerning body to which are added Six Lessons to the Professors of Mathematicks of the Institution of Sir Henry Savile, in the University of Oxford.* London, 1656.

———. *The English Works of Thomas Hobbes of Malmesbury.* Edited by William Molesworth. 11 vols. London: 1839–1845.

———. *Opera Philosophica.* Edited by William Molesworth. 5 vols. London, 1839–1845.

Hofmann, Joseph E. *Die Entwicklungsgeschichte der Leibnizschen Mathematik während des Aufenthaltes in Paris (1672–1676).* Munich: Leibniz Verlag, 1949.

———. *Leibniz in Paris, 1672–1676: His Growth to Mathematical Maturity.* New York: Cambridge University Press, 1974.

———. *Leibniz' mathematische Studien in Paris.* Berlin: W. de Gruyter, 1948.

Husserl, Edmund. *La crise des sciences européennes et la phénoménologie transcendentale.* Translated by Gérard Granel. Paris: Gallimard, 1976.

Huygens, Christiaan. *Christiaan Huygens' "The Pendulum Clock, or, Geometrical Demonstrations concerning the Motion of Pendula as Applied to Clocks."* Translated, with notes, by Richard J. Blackwell. Introduction by H. J. M. Bos. Ames: Iowa State University Press, 1986.

———. *De motu corporum ex percussione* (Leyden, 1703). In *Opuscula Posthuma* and also in *Oeuvres complètes,* vol. 16.

———. Manuscripts. University of Leyden Library.

———. *Oeuvres complètes de Christiaan Huygens.* 22 vols. The Hague: Société Hollandaise des Sciences, 1888–1950.

————. *Opuscula Posthuma*. Edited by B. de Volder and B. Fullenius. Leyden, 1703.

————. "Sur les Règles du mouvement dans la rencontre des corps." *Journal des Sçavans* (18 March 1669).

Jouguet, Émile. *Lectures de mécanique*. Paris: Gauthier-Villars, 1924.

Kepler, Johannes. *Ad Vitellionem Paralipomena*. Prague, 1604.

————. *Epitome astronomiae Copernicae*. Linz, 1617, 1621.

————. *Gesammelte Werke*. Edited by Walther van Dyck, Max Caspar, and F. Hammer. Munich: C. H. Beck, 1937–.

Kockelmans, Joseph J. *Heidegger and Science*. Washington, D.C.: Center for Advanced Research in Phenomenology and University Press of America, 1985.

Koyré, Alexandre. *The Astronomical Revolution: Copernicus, Kepler, Borelli*. Translated by R. E. W. Watson. Ithaca: Cornell University Press, 1973.

————. *Chute des corps et mouvement de la terre de Kepler à Newton*. Paris: Vrin, 1973.

————. *Études d'histoire de la pensée philosophique*. Paris: Colin, 1961. Reprint, Paris: Gallimard, 1971.

————. *Études d'histoire de la pensée scientifique*. Paris: Presses Universitaires de France, 1966. Reprint, Paris: Gallimard, 1973.

————. *Études galiléennes*. Paris: Hermann, 1966.

————. *Études newtoniennes*. Paris: Gallimard, 1968.

————. *From the Closed World to the Infinite Universe*. Baltimore: Johns Hopkins University Press, 1957.

————. *Galileo Studies*. Translated by John Mepham. Atlantic Highlands, N.J.: Humanities Press, 1978.

Kuhn, Thomas S. *The Copernican Revolution: Planetary Astronomy in the Development of Western Thought*. Cambridge: Harvard University Press, 1957.

Lagrange, Joseph-Louis. *Mécanique analitique*. Paris, 1788.

————. *Mécanique analytique*. Complete edition containing the notes to the 3d ed. (1853–1855), revised, corrected, and annotated by Joseph Bertrand, and to the 4th ed. (1888), edited by Gaston Darboux. 2 vols. Paris: Éditions A. Blanchard, 1965.

————. *Mécanique analytique*. 3d ed. Paris, 1853–1855.

————. *Théorie des fonctions analytiques, contenant les principes du calcul différentiel, dégagés de toute considération d'infiniment petits, d'évanouissants, de limites et de fluxions, et réduits à l'analyse algébrique des quantités finies*. Paris, 1797.

La Harpe, Jacqueline de. "Des inédits de Fontenelle: Sa correspondence avec J.-P. Crousaz." *Revue historique vaudoise* (June 1954): 90–108.

Laplace, Pierre-Simon. *Exposition du Système du Monde*. In *Oeuvres complètes de Laplace*, vol. 6. 6th ed. Paris, 1835.

Leibniz, Gottfried Wilhelm. *Animadversiones in partem generalem Principiorum Cartesianorum* (1692). In *Die Philosophischen Schriften von Leibniz*, 4:350–392.

————. *Briefwechsel zwischen Leibniz und Christian Wolf aus den Handschriften der Koeniglichen Bibliothek zu Hannover*. Edited by C. I. Gerhardt. Halle, 1860.

————. *Der Briefwechsel des Gottfried Wilhelm Leibniz in der Königlichen Öffentlichen Bibliothek zu Hannover*, edited by Eduard Bodemann. Hanover, 1889.

————. *De Corporum Concursu* (January–February 1678). In Leibniz-Handschriften, Mathematik, vol. 35. Niedersächsische Landesbibliothek, Hanover.

————. "De geometria recondita et analysi indivisibilium atque infinitorum." *Acta Eruditorum* (June 1686): 292–300.

————. "Essay de Dynamique sur les lois du mouvement" (1699). In *Leibnizens mathematische Schriften*, 6:215ff.

————. *Hypothesis physica nova qua phaenomenorum naturae pleororumque causae ab unico quodam universali motu, in globo nostro supposito, neque tychonicis, neque copernicanis aspernando, repetuntur*. Mayence, 1671.

————. *The Leibniz-Arnauld Correspondence*. Edited and translated by H. T. Mason. Manchester: Manchester University Press, 1967.

————. *Leibnizens mathematische Schriften*. Edited by C. I. Gerhardt. 7 vols. Berlin and Halle, 1849–1863. Reprint, Hildesheim: Georg Olms, 1960–1961.

————. "Letter of Mr. Leibniz on a General Principle Useful in Explaining the Laws of Nature through a Consideration of the Divine Wisdom; to Serve as a Reply to the Response of the Rev. Father Malebranche" (July 1687). In *Philosophical Papers and Letters*, 351–354.

————. *Die Leibniz-Handschriften der Königlichen Öffentlichen Bibliothek zu Hannover*, edited by Eduard Bodemann. Hanover and Leipzig, 1867 and 1895. Reprint, Hildesheim: Georg Olms, 1966.

————. *Lettres et opuscules inédits de Leibniz*. Edited by L.-A. Foucher de Careil. Paris, 1854.

————. *Naissance du calcul différentiel: 26 articles des "Acta Eruditorum."* Translated, with introduction and notes, by Marc Parmentier. Paris: Vrin, 1989.

————. *New Essays on Human Understanding*. Translated and edited by Peter Remnant and Jonathan Bennett. Cambridge: Cambridge University Press, 1981.

————. "Nova methodus pro maximis et minimis, itemque tangentibus, quae nec fractas, nec irrationales quantitates moratur, et singulare pro illis calculi genus." *Acta Eruditorum* (October 1684): 467–473.

————. *Philosophical Papers and Letters*. Translated and edited by Leroy E. Loemker. 2d ed. Dordrecht: D. Reidel, 1976.

————. *Die Philosophischen Schriften von Leibniz*. Edited by C. I. Gerhardt. 7 vols. Berlin and Halle: 1875–1890; reprint, Hildesheim: Georg Olms, 1960–1961.

————. "Réponse de M. L. à la Remarque de M. l'Abbé D.C. contenue dans l'article 1. de ces Nouvelles, mois de juin 1687, où il prétend soutenir une loi de la Nature avancée par M. Descartes." *Nouvelles de la République des Lettres* (September 1687): 952–956.

————. *Sämtliche Schriften und Briefe*. 25 vols. Berlin: Akademie Verlag, 1923–.

————. "Specimen Dynamicum" (1695). Edited by Hans Gunter Dosch, Glenn W. Most, and Enno Rudolph. Hamburg: F. Meiner Verlag, 1982.

————. "Tentamen Anagogicum: Essay anagogique dans la recherche des causes." In *Die Philosophischen Schriften,* 7:270–279.

————. "Unicum opticae, catoptricae et dioptricae principium." *Acta Eruditorum* (June 1682): 185–190.

Le Seur, Thomas. *Philosophia Naturalis Principia Mathematica auctore Isaaco Newtono, Eq. Aurato: Perpetuis Commentariis Illustrata.* In collaboration with F. Jacquier. 4 vols. Geneva, 1739–1742.

Levy, Tony. *Figures de l'infini: Les mathématiques au miroir des cultures.* Paris: Seuil, 1987.

l'Hospital, Guillaume F. A. de. *Analyse des infiniment petits pour l'intelligence des lignes courbes.* Paris, 1696.

————. *The method of fluxions both direct and inverse.* Translated by E. Stone. London, 1730.

Mach, Ernst. *The Science of Mechanics: A Critical and Historical Account of Its Development.* Translated by Thomas J. McCormack. 6th ed. (with revisions through the 9th German ed.). Lasalle, Ill.: Open Court, 1960. 3d paperback ed., 1974.

Malebranche, Nicolas. *De la Recherche de la Vérité.* Paris, 1675.

————. *Oeuvres complètes de Malebranche.* Edited by André Robinet. 21 vols. Paris: Vrin, 1958–1970. 2d ed., 1967–1978.

Maltese, Giulio. *La storia di "F = ma": La seconda legge del moto nel XVIII secolo.* Florence: Olschki, 1992.

Mariotte, Edme. *Essai de logique* (Paris, 1678). Edited by Guy Picolet in collaboration with Alain Gabbey. Paris: Librairie Fayard, 1992.

————. *Oeuvres de Mr. Mariotte.* 2 vols. Leyden, 1717.

————. *Traitté de la percussion ou chocq des corps: Dans lequel les principales Règles du mouvement contraires à celles de Mr. Descartes, & quelques autres Modernes ont voulu établir, sont démonstrées par leurs véritables causes.* Paris, 1673.

Merleau-Ponty, Jacques. *Leçons sur la genèse des théories physiques: Galilée, Ampère, Einstein.* Paris: Vrin, 1974.

Mersenne, Marin. *Correspondance du P. Marin Mersenne.* Edited and annotated by Cornelis de Waard, Bernard Rochot, and Armand Beaulieu. 15 vols. Paris: Centre National de la Recherche Scientifique, 1932–1983.

————. *Harmonie universelle contenant la théorie et la pratique de la musique.* Paris, 1636.

Meyerson, Emile. *De l'explication dans les sciences.* 2 vols. Paris: Payot, 1921.

————. *Essais.* Paris: Vrin, 1936.

————. *Identité et realité.* 5th ed. Paris: Vrin, 1951.

Milhaud, Gaston. *Essai sur les conditions et les limites de la certitude logique.* 4th ed. Paris: Alcan, 1924.

————. *Études sur la pensée scientifique chez les Grecs et chez les modernes.* Paris: Société Française d'Imprimerie et de Librairie, 1906.

————. *Nouvelles études sur l'histoire de la pensée scientifique.* Paris: Alcan, 1911.

————. *Le rationnel: Étude complémentaire à l'essai sur la certitude logique.* 2d ed. Paris: Alcan, 1926.

Milliet de Chales, Claude-François. *Cursus seu mundus mathematicus*. 3 vols. Lyons, 1674.

———. *Traité du mouvement local et du ressort*. Lyons, 1682.

Montucla, J.-F. (Jean-Étienne). *Histoire des mathématiques*. 2d ed. 4 vols. Paris, 1799–1802.

More, Henry. *Enchiridion Metaphysicum, sive, De rebus incorporeis succincta & luculenta dissertatio*. London, 1671.

Moscovici, Serge. *Jean-Baptiste Baliani, disciple et critique de Galilée: L'expérience du mouvement*. Paris: Hermann, 1967.

Mouy, Paul. *Le développement de la physique cartésienne, 1646–1712*. Paris: Vrin, 1934. Reprint, New York: Arno Press, 1981.

Nardi, Antonio. "La quadratura della velocità: Galileo, Mersenne, La Tradizione." *Nuncius* 3 (1988): 27–64.

Nelson, Edward. "Internal set theory: A new approach to NSA." *Bulletin of the American Mathematical Society* 83 (1977): 1165–1198.

Newton, Isaac. *The Correspondance of Isaac Newton*. Edited by H. W. Turnbull, J. F. Scott, A. R. Hall, and L. Tilling. 7 vols. Cambridge: Cambridge University Press, 1959–1977.

———. *De gravitatione* (1665–1670). Manuscript Add 4003, Cambridge University Library. In *Unpublished Scientific Papers of Isaac Newton*, edited by A. R. Hall and Mary Boas-Hall. Cambridge: Cambridge University Press, 1962.

———. *De gravitatione*. Reprinted with introduction by François De Gandt (including a translation of *De motu*). Paris: Gallimard, 1995.

———. *De la gravitation*. Translated, with notes and introduction, by Marie-Françoise Biarnais. Paris: Les Belles Lettres, 1985.

———. *Isaac Newton's Papers and Letters on Natural Philosophy*. Edited by I. Bernard Cohen with the assistance of Robert E. Schofield. Cambridge: Harvard University Press, 1958.

———. *The Mathematical Papers of Isaac Newton*. Edited by D. T. Whiteside. 8 vols. Cambridge: Cambridge University Press, 1967–1981.

———. *La méthode des fluxions et des suites infinies*. Translated by George-Louis Leclerc, Comte de Buffon. Paris, 1740. Reprint, Paris: Blanchard, 1966.

———. *Philosophiae Naturalis Principia Mathematica*. London, 1687.

———. *Philosophiae Naturalis Principia Mathematica: The Third Edition (1726) with Variant Readings*. Edited by Alexandre Koyré and I. Bernard Cohen. Cambridge: Cambridge University Press, 1972.

———. *Principes mathématiques de la philosophie naturelle*. Translated by Gabrielle-Émilie de Breteuil, Marquise du Chastelet. Original manuscript. Paris, 10 September 1749. Bibliothèque National Fonds Français, 12266–12268.

———. *Les principes mathématiques de la philosophie naturelle*. New partial translation, with afterword and bibliography, by Marie-Françoise Biarnais. Paris: Centre de Documentation Sciences Humaines (CNRS), 1982. Reprint, Paris: Christian Bourgois, 1985.

———. *Principes mathématiques de la philosophie naturelle*. Translated by Gabrielle-Émilie de Breteuil, Marquise du Chastelet, and revised by Alexis-Claude Clairaut. 2 vols. Paris, 1756–1759. Reprint, Paris: Blanchard, 1966; Gabay, 1989.

————. *Sir Isaac Newton's Mathematical Principles of Natural Philosophy and His System of the World*. Translated by Andrew Motte and revised by Florian Cajori. Berkeley and Los Angeles: University of California Press, 1934.

Niderst, Alain. *Fontenelle à la recherche de lui-même (1657– 1702)*. Paris: Nizet, 1972.

Nieuwentijt, Bernard. *Analysis infinitorum seu curvilineorum proprietates ex polygonorum deductae*. Amsterdam, 1695.

————. *Considerationes circa analyseos ad quantitates infinite parvas applicatae principia, et calculi differentialis usum in resolvendis problematibus geometricis*. Amsterdam, 1694.

————. *Considerationes secundae circa calculi differentialis principia, et responsio ad virum nobilissimum G. G. Leibnitium*. Amsterdam, 1696.

Palter, Robert, ed. *The "Annus Mirabilis" of Sir Isaac Newton, 1666–1966*. Cambridge: MIT Press, 1970.

Panofsky, Erwin. *Galileo as a Critic of the Arts*. The Hague: M. Nijhoff, 1954.

Pascal, Blaise. *De l'esprit géométrique* (Paris, 1657/58). In *Great Shorter Works*.

————. *Great Shorter Works of Pascal*. Translated by Émile Cailliet and John C. Blankenagel. Philadelphia: Westminster Press, 1948.

————. *Oeuvres complètes*. Edited by Louis Lafuma. Paris: Seuil, 1963.

————. *Pensées*. Translated by A. J. Krailsheimer. Harmondsworth: Penguin, 1966.

Ptolemy [Claudius Ptolemaeus]. *"L'Optique" de Claude Ptolémée, dans la version latine d'après l'arabe de l'émir Eugène de Sicile*. Edited, with notes and French translation, by Albert Lejeune. Louvain: Bibliothèque de l'Université, 1956. Reprint, Leyden: E. J. Brill, 1989.

Roberval, G. Personne de. *Élémens de géométrie* (1675). Archives of the Royal Academy of Sciences.

Robinet, André. *Architechtonique disjonctive: Automates systémiques et idéalité transcendentale dans l'oeuvre de G. W. Leibniz*. Paris: Vrin, 1986.

————. "Le group malebranchiste introducteur du calcul infinitésimal en France." *Revue d'histoire des sciences* 13, no. 3 (1960): 287–308.

————. *G. W. Leibniz: Iter Italicum (March 1689–March 1690)*. Florence: Olschki, 1988.

————. *G. W. Leibniz: Phoranomus Seu de Potentia et Legibus Naturae*. Florence: Olschki, 1991.

————. *Malebranche de l'Académie des Sciences: L'oeuvre scientifique, 1674–1715*. Paris: Vrin, 1970.

————. *Malebranche et Leibniz: Relations personelles présentées avec les textes complets des auteurs et de leurs correspondants revus, corrigés et inédits*. Paris: Vrin, 1955.

Robinson, Abraham. *Non-standard Analysis*. Amsterdam: North-Holland, 1966.

Rolle, Michel. "Troisième remarque sur les principes de la géométrie des inf. petits." Registres manuscrits des procès-verbaux des séances de l'Académie Royale des Sciences de Paris, vol. 20, fols. 95r–101r (March 1701).

Saurin, Joseph. "Démonstration des Théorèmes que M. Hugens a proposez dans son Traité de la pendule sur la force centrifuge des corps mus circulairement." *Journal des Trévoux* (1702): 27–60.

Schafheitlin, Paul. "Johann Bernoullis Differentialrechnung." *Verhandlungen der Naturforschenden Gesellschaft* 33 (1920–1921): 230–235.

———. "Johann Bernoulli lectiones de calculo differentialum." *Verhandlungen der Naturforschenden Gesellschaft* 34 (1922): 1–31.

Scylla, Edith D. *The Oxford Calculators and the Mathematics of Motion, 1320–1350: Physics and Measurements by Latitudes.* New York: Garland, 1991.

Serres, Michel. *Le système de Leibniz et ses modèles mathématiques.* Paris: Presses Universitaires, 1968. Reprint, 1982.

Shea, William R., ed. *Nature Mathematized: Historical and Philosophical Case Studies in Classical Modern Natural Philosophy.* Dordrecht: D. Reidel, 1983.

———. *La révolution galiléenne: De la lunette au système du monde.* Translated by François De Gandt. Paris: Seuil, 1992.

Simon, Gérard. *Kepler: Astronome astrologique.* Paris: Gallimard, 1979.

Struik, D. J., ed. *A Source Book in Mathematics, 1200–1800.* Princeton: Princeton University Press, 1986; 1st edition, 1969.

Szabo, Istvan. *Geschichte der Mechanischen Prinzipien und ihrer wichtigsten Anwendungen.* 3d ed. Basel: Birkhaüser, 1987.

Tacquet, André. *Opera mathematica R. P. Andreae Tacquet, [. . .] demonstrata et propugnata a Simone Laurentio Veterani, ex comitibus Monti Calvi, in Collegio Societatis Jesu, Lovanii, anno 1668.* Antwerp, 1669. Reprinted in 1707 in 2 vols.

Timoshenko, Stephen P. *History of Strength of Materials.* New York: McGraw-Hill, 1953. Reprint, New York: Dover Publications, 1983.

Torricelli, Evangelista. *De motu gravium naturaliter descendentium et projectorum.* Florence, 1644. In *Opere,* vol. 2.

———. *Opere di Evangelista Torricelli.* Edited by Gino Loria and Giuseppe Vassura. 4 vols. Faenza: G. Montanavi, 1919–1944.

———. *Opera geometrica.* Florence, 1644.

———. *Traicté du Mouvement des eaux d'Évangeliste Torricelli mathématicien du Grand Duc de Toscane: Tiré du Traité du mesme autheur, du Mouvement des corps pesans qui descendent naturellement et qui sont jetez.* Translated by P. Saporta. Castres, 1664.

Trublet, Nicolas Charles Joseph. *Mémoires pour servir à l'histoire de la vie et des ouvrages de Fontenelle tirés du Mercure de France, 1756, 1757, et 1758, par l'abbé Trublet, seconde édition corrigée et augmentée.* Amsterdam, 1759.

Truesdell, Clifford. *Essays in the History of Mechanics.* Berlin: Springer-Verlag, 1968.

———. *The Rational Mechanics of Flexible or Elastic Bodies, 1638–1788: Introduction to "Leonhardi Euleri Opera Omnia," Vols. X and XI, Seriei Secundae.* Zurich: Orell Fussli, 1960.

Varignon, Pierre. "Application de la Regle generale des vitesses variées, comme on voudra, aux mouvemens par toutes sortes de courbes, tant mecaniques que geometriques. D'ou l'on deduit encore une nouvelle maniere de démontrer les chutes isochrones dans la cycloïde renversée." Registres manuscrits des procès-verbaux des séances de l'Académie Royale des Sciences de Paris, vol. 17, fols. 387r–391v (6 September 1698).

———. "Courbe isochrone: Le long de laquelle descendent d'une vitesse uniforme par rapport à l'horizon, en sorte qu'ils s'en approchent également en temps égaux." Registres manuscrits des procès-verbaux des séances de l'Académie Royale des Sciences de Paris, vol. 14, fols. 135v–136r (30 July 1695).

———. "Démonstration de six manières différentes de trouver les rayons des développées, lors même que les ordonnées des courbes qu'elles engendrent, concourent en quelque point que ce soit et par conséquent aussi pour le cas où elles sont parallèles." Archives of the Royal Academy of Sciences (Paris), Pochettes de Séances, 27 November 1694, 3 pp.

———. "Démonstration générale de l'arithmétique des infinis ou de la géométrie des indivisibles." Archives of the Royal Academy of Sciences (Paris), Pochettes de Séances, 2 January 1694, 4 pp.

———. "Du déroulement des spirales de tous les genres, où l'on fait voir qu'elles se déroulent toutes en paraboles d'un degré seulement plus haut que leur, avec une méthode générale pour toutes ces sortes de déroulement." Registres manuscrits des procès-verbaux des séances de l'Académie Royale des Sciences de Paris, vol. 14, fols. 192r–193r (12 November 1695).

———. "Du mouvement des eaux, ou d'autres liqueurs quelconques de pesanteurs spécifiques à discrétion; de leurs vitesses, de leurs dépenses par telles ouvertures ou sections qu'on voudra; de leurs hauteurs au-dessus de ces ouvertures, des durées de leurs écoulemens etc." Histoire de l'Académie Royale des Sciences (Mémoires), year 1703 (1705): 238–261.

———. "Du mouvement en générale par toutes sortes de courbes; et des forces centrales, tant centrifuges que centripètes, nécessaires aux corps qui les décrivent." Registres manuscrits des procès-verbaux des séances de l'Académie Royale des Sciences de Paris, vol. 19, fols. 133v–141v (31 March 1700); also in Histoire de l'Académie Royale des Sciences (Mémoires), year 1700 (1703): 83–101.

———. "Liste des oeuvres de Pierre Varignon." Edited by J. O. Fleckenstein, P. Costabel, J. Peiffer, and B. Bilodeau. In Jean Bernoulli, Briefwechsel, 2:387–408.

———. "Manière générale de déterminer les forces, les vitesses, les espaces, et les temps, une seule de ces quatre choses étant données dans toutes sortes de mouvemens rectilignes variés à discrétion." Registres manuscrits des procès-verbaux des séances de l'Académie Royale des Sciences de Paris, vol. 19, fols. 31r–37r (30 January 1700); also in Histoire de l'Académie Royale des Sciences (Mémoires), year 1700 (1703): 22–27.

———. "Manière générale de trouver les tangentes spirales de tous les genres et de tant de révolutions qu'on voudra avec leurs quadratures indéfinies." Archives of the Royal Academy of Sciences (Paris), Pochettes de Séances, 11 December 1694, 5 pp.

———. Nouvelle mécanique ou statique dont le projet fut donné en 1687: Ouvrage posthume de M. Varignon. Paris, 1725.

———. "Rectification et quadrature indéfinies des cycloïdes à bases circulaires, quelque distance qu'on suppose entre leur point décrivant et le centre de leur cercle mobile." Registres manuscrits des procès-verbaux des séances de l'Académie Royale des Sciences de Paris, vol. 14, fols. 181r–184r (3 September 1695).

———. "Réfutation du sentiment du P. Guldin et de MM. Wallis et Sturmius sur la longueur de la spirale d'Archimède." Archives of the Royal Academy of Sciences (Paris), Pochettes de Séances, 13 March 1694, 2 pp.

———. "Regle generale pour toutes sortes de mouvemens de vitesses quelconques variées a discretion." Registres manuscrits des procès-verbaux des séances de l'Académie Royale des Sciences de Paris, vol. 17, fols. 297v–305r (5 July 1698).

———. "Règles du mouvement en général." In *Mémoires de l'Académie Royale des Sciences depuis 1666 jusqu'à 1699,* 10:225–233.

———. *Traité du mouvement et de la mesure des eaux coulantes et jalissantes.* Paris, 1725.

Viète, François. *Isagoge in artem analyticem.* Tours, 1591.

Wallis, John. *Arithmetica Infinitorum, sive, Nova methodus inquirendi in curvilineorum quadraturam, aliaq[ue] difficiliora matheseos problemata.* Oxford, 1655.

———. *De sectionibus conicis: novo methodo expositis tractatus.* London, 1655.

———. *Elenchus geometriae Hobbianae sive, geometricorum, quae in ipsius elementis philosophiae, a Thoma Hobbes Malmesburiensi proferuntur.* Oxford, 1655.

———. *Mechanica sive de motu tractatus geometricus.* London, 1670–1671.

———. *Opera Mathematica.* 3 vols. Oxford, 1695–1699.

———. *A Treatise of Algebra, both historical and practical.* London, 1684–1685.

Wallis, Peter John, and Ruth Wallis. *Newton and Newtoniana, 1672–1975: A Bibliography.* Folkstone: Dawson, 1977.

Westfall, Richard S. *Force in Newton's Physics.* London: Macdonald, 1971.

———. *Never at Rest: A Biography of Isaac Newton.* Cambridge: Cambridge University Press, 1980.

Weyl, Hermann. *Philosophy of Mathematics and Natural Science.* Princeton: Princeton University Press, 1949.

Whiteside, D. T. *The Mathematical Principles Underlying Newton's "Principia Mathematica."* Glasgow: University of Glasgow, 1970.

———. "Patterns of Mathematical Thought in the Later Seventeenth Century." *Archive for History of Exact Sciences* 1 (1961): 179–388.

Wisan, W. L. "The New Science of Motion: A Study of Galileo's *De Motu Locali.*" *Archives for History of Exact Sciences* 13 (1974): 103–306.

Wronski, Hoené. *Philosophie critique, fondée sur le premier principe du savoir humain.* Marseilles, 1803.

———. *La philosophie de l'infini: Contenant des contre-refléxions sur la métaphysique du calcul infinitesimal.* Paris, 1814.

Wurtz, J.-P. "La naissance du calcul différentiel et le problème du statut des infiniment petits." In *L'Analyse non standard: Recherches historiques et philosophiques.* Working Paper 265 of the Centre National de la Recherche Scientifique. Paris, 1987.

Yoder, Joella G. *Unrolling Time: Christiaan Huygens and the Mathematization of Nature.* Cambridge: Cambridge University Press, 1988.

Youschkevitch, A. P. "Comparaison des conceptions de Leibniz et de Newton sur le calcul infinitesimal." *Studia Leibnitiana,* Supplementa 17 (1978): 69–80.

———. "The Concept of Function up to the Middle of the 19th Century." *Archive for History of Exact Sciences* 15 (1976–1977): 37–85.

Index